SYDNEY BRENNER

A BIOGRAPHY

ALSO FROM COLD SPRING HARBOR LABORATORY PRESS

SYDNEY BRENNER

A BIOGRAPHY

Errol C. Friedberg

University of Texas Southwestern Medical Center at Dallas

Dallas, Texas

 COLD SPRING HARBOR LABORATORY PRESS

Cold Spring Harbor, New York • www.cshlpress.com

SYDNEY BRENNER: A BIOGRAPHY

© 2010 by Cold Spring Harbor Laboratory Press, Cold Spring Harbor, New York
All rights reserved
Printed in the United States of America

Publisher & Acquisitions Editor	John Inglis
Director Development, Marketing, & Sales	Jan Argentine
Developmental Editor	Kaaren Janssen
Project Coordinator	Maryliz Dickerson
Permissions Administrator	Carol Brown
Production Manager	Denise Weiss
Production Editor	Mala Mazzullo
Typesetter	Techset Composition Limited
Interior Book Designer	Denise Weiss
Cover Designer	Ed Atkeson

Front Cover: Photograph of Sydney Brenner taken by Jane Bown in the 1980s.
(Courtesy of Sydney Brenner.)

Library of Congress Cataloging-in-Publication Data

Friedberg, Errol C.
  Sydney Brenner : a biography / Errol C. Friedberg.
    p. cm.
  Includes bibliographical references and index.
  ISBN 978-0-87969-947-5 (hardcover : alk. paper)
  1. Brenner, Sydney. 2. Biologists—South Africa—Biography. I. Title.

  QH31.B82F75 2011
  570.92—dc22
  [B]

2010017671

10 9 8 7 6 5 4 3 2

All World Wide Web addresses are accurate to the best of our knowledge at the time of printing.

Authorization to photocopy items for internal or personal use, or the internal or personal use of specific clients, is granted by Cold Spring Harbor Laboratory Press, provided that the appropriate fee is paid directly to the Copyright Clearance Center (CCC). Write or call CCC at 222 Rosewood Drive, Danvers, MA 01923 (978-750-8400) for information about fees and regulations. Prior to photocopying items for educational classroom use, contact CCC at the above address. Additional information on CCC can be obtained at CCC Online at http://www.copyright.com/.

All Cold Spring Harbor Laboratory Press publications may be ordered directly from Cold Spring Harbor Laboratory Press, 500 Sunnyside Boulevard, Woodbury, New York 11797-2924. Phone: 1-800-843-4388 in Continental U.S. and Canada. All other locations: (516) 422-4100. FAX: (516) 422-4097. E-mail: cshpress@cshl.edu. For a complete catalog of Cold Spring Harbor Laboratory Press publications, visit our World Wide Web Site http://www.cshlpress.com/.

Men of genius do not excel in any profession because they labor in it, but they labor in it because they excel.

WILLIAM HAZLITT
English essayist (1778–1830)

For Rhonda
&
In memory of May Brenner
(1921–2010)

Contents

PART 4
Complex Organisms

PART 5
Life Outside the Laboratory

See photo section between pages 110 and 111.

Preface

In the spring of 1953 at the Cavendish Laboratory, Cambridge, James Watson and Francis Crick presented a model of the structure of DNA. This glimpse of the gene that lent Watson and Crick such historic status marked the founding of molecular biology, a discipline characterized by the integration of biochemistry and genetics that has since dominated much of modern biology. The years following are often referred to as a "golden age," an era comparable in impact to the revolution in physics that transpired earlier in the 20th century.

Sydney Brenner, an enthusiastic and talented 22-year-old biologist from South Africa, was one of the earliest visitors to view the newly unveiled DNA structure. Brenner, then a graduate student in the Department of Physical Chemistry at Oxford, made a striking impression on Crick, who was determined to recruit the young South African to Cambridge. His efforts yielded extraordinary dividends. In the following years, the two deciphered many of the elements of gene function in a breathtaking series of contributions that surely marks one of the most creative periods in the history of biology. Their intellectual partnership dissolved when Brenner sought new research horizons and alighted on the nematode *Caenorhabditis elegans*, a tiny worm that became a celebrated model organism for probing the complexities of life in multicellular organisms. The 2002 Nobel Prize in Medicine or Physiology recognized the importance of this contribution.

Sydney Brenner is widely regarded as one of the leading molecular biologists of the twentieth century. But the effort of documenting his life is not restricted to recounting his contributions in the research laboratory. He was a significant influence in moderating the frenetic debates on recombinant DNA technology in the mid-1970s, and in helping to orchestrate Britain's involvement in the Human Genome Project in the early 1990s. And for about a decade he directed the Laboratory of Molecular Biology in Cambridge (which replaced the crowded and dilapidated Cavendish Laboratory) with its excellent scientific staff—many Nobel Laureates.

Rules established by the Medical Research Council in the United Kingdom required that Brenner formally retire as a paid scientist in 1992, when he reached the age of 65. But, at the time of this writing, Brenner is as active as ever scientifically, showing no signs of slowing down. In his later years he profoundly influenced the emergence of cutting-edge biomedical research in Singapore, one of the Asian tigers seeking to break into the front ranks of molecular biology and biotechnology. He has since inspired and lent his organizational skills to restructuring the sociology of molecular biology in Japan and to help guide the future of the Janelia Farm campus of the Howard Hughes Medical Institute. All these efforts were undertaken while Brenner actively continued to guide diverse scientific projects in other parts of the world and to promote efforts in the biotechnology sector.

Brenner's single-minded passion for biology has long dominated his waking—and presumably more than a few of his sleeping—hours, leaving him little time for other pursuits. But aside from his scientific contributions, which remain undiminished, Brenner possesses a broad intellect that embraces more than a superficial knowledge of the arts and history, and his talent as a raconteur is widely celebrated. He has dazzled, amused, and offended countless audiences with his wit and ironic humor; his iconoclastic views on ideas related to the exploration of life on this planet (and on planets yet unseen); and his general disdain of authority and dogma. He is, in fact, the proverbial enfant terrible.

An inveterate talker, Brenner can (and usually does) dominate any conversation of which he is part. But ironically, he is very much a loner, far preferring to think about and execute scientific experiments than to cavort with friends and acquaintances. As is evident among the photographs in this volume, he tends to set himself apart in group situations, and his countenance sometimes reflects the utter boredom associated with time away from the laboratory, the library, or his desk.

This biography begins with Brenner's humble beginnings as the child of an immigrant cobbler father and homemaker mother in the town of Germiston, South Africa. It then follows his educational path, from his years as a medical student at the University of the Witwatersrand in Johannesburg, South Africa, through his sojourn at Oxford University where he acquired a second doctoral degree, to his long association with Cambridge University and his wandering career as a "retired" scientist. Much of the book is based on information from personal interviews with Brenner and with a number of his former and current scientific associates, friends, and relatives. My efforts were also helped considerably by a 15-hour videotaped interview by

Lewis Wolpert in 1994, that I converted to a chronologically comprehensive text, entitled *Sydney Brenner: My Life in Science*, published in 2001 by Bio-Med Central. All unreferenced quotations from Brenner are from this source (regrettably now out of print). Readers should be aware that direct quotations from Brenner and others of British or British colonial origin use traditional English spelling. I have also quoted (accurately, I hope) from my interviews with others.

Sydney Brenner is among the very few key individuals to foster the early development of the discipline of molecular biology. Clearly history will determine how that period should be viewed in the context of preceding and succeeding events in the world of biology. This book therefore is neither offered as a definitive documentation of Sydney Brenner's life, nor as an attempt to establish his place in the annals of science. It is, rather, my hope that it will provide a useful foundation for more detailed and analytical contributions by future scholars and commentators.

Evolution is a central topic among Sydney's many eclectic scientific interests. For this reason alone I am delighted that this work was essentially completed in 2009, a year that celebrates the 200th anniversary of the birth of Charles Darwin. Darwin and Gregor Mendel rank high on Brenner's very short list of scientific heroes.

ERROL C. FRIEDBERG
February 2010

Acknowledgments

THIS BOOK WOULD NOT HAVE SEEN THE LIGHT OF DAY without the help of many individuals to whom I am deeply indebted. First and foremost, I thank Sydney Brenner for unrestricted access that consumed many hours of his time and energy over the course of nearly four years, not to mention his hospitality in inviting me into his various abodes. Sydney never asked to read the manuscript prior to publication; nor did he restrict or censor my writing in the least. But I am enormously grateful to him, too, for taking the time to read a near final draft to correct historical inaccuracies. I thank Sam Aparicio, Jonathan Balkind, David Baltimore, Sara Bancroft, John Baughman, the late Seymour Benzer, Paul Berg, Walter Bodmer, Carla Brenner, the late May Brenner, Roger Brent, Mark Bretscher, Neal Copeland, the late Odile Crick, Manfred Eigen, Sam Eletr, Phyllis Finn, Stanley Glasser, Philip Goelet, Richard Henderson, Robert Horvitz, Hugh Huxley, Charles Isaacson, François Jacob, Kim Janda, Nancy Jenkins, Jonathan Karn, Aaron Klug, Peter Lachmann, David Lane, Richard Lerner, Solly Levin, Louie Lim, Bronwen Loder, Matthew Meselson, Jeffrey Miller, Michael Neuberger, Peter Newmark, Alice Orgel, the late Leslie Orgel, Keith Peters, Lim Pin, Gerry Rubin, Fred Sanger, Terry Sejnowski, John Sulston, Chris Tan, Phillip Tobias, Vitek Tracz, Byrappa Venkatesh, Paul Wassarman, James Watson, Muriel Wigby, Lewis Wolpert, Charles Yanofsky, and Philip Yeo for taped interviews. These are now stored in the Sydney Brenner Collection at the Cold Spring Harbor Laboratory Library and Archives.

Special thanks are owed to my sister Pearl Benater, Diego Castrillon, Andrew Friedberg, Sylvia Friedberg, Bob Horvitz, Trefor Jenkins, Jonathan Karn, Bronwen Loder, Ethel Moustacchi, Michael Neuberger, Maria del Pilar Martin, Nancy Schneider, Terry Sejnowski, Paul Wassarman, Keith Wharton, and Wei Yang for critical reading of one or more drafts. My wife Rhonda diligently read and critiqued more versions of the manuscript than she cares to admit and provided frequently needed support and encouragement, even at times when she thought that my attention to this labor was a little beyond the pale. Sincere thanks to Ludmilla Pollock, Director of the

Cold Spring Harbor Laboratory Library and Archives for facilitating access to Brenner's papers. Thanks also to Mila's dedicated staff of archivists, notably Clare Clark, Anthony Dellureficio, and Patricia Novak, for guiding me through this massive collection, none of which was formally archived at the time of my research.

My assistant Meredith Thomas maintained cabinets teeming with files, organized countless trips to far-flung corners of the world, and generally kept me organized throughout my writing, often executed in the midst of far too many other activities. Thanks also to Mark Smith for invaluable assistance in rendering electronic images of photographs, to Vicky Hernandez who made sure I didn't overspend my budget, and to Sara Bancroft, Sydney's longtime and long distance assistant, for tracking his movements around the world. To my knowledge Sara is the only person on this planet who knows which nook of the earth Sydney Brenner is in at any given moment; though sometimes his movements flummox even her.

I am enormously grateful to John Inglis, Director of Cold Spring Harbor Laboratory Press, for undertaking publication of this book. Thanks also to Kaaren Janssen and Alex Gann for their excellent editing and attempts to rein in the hyperbole and excessive use of needless adjectives; and to Maryliz Dickerson whose efforts were indispensable to keeping us organized and moving ahead. I thank also Jan Argentine, Mala Mazzullo, and Denise Weiss at the Press for their critical support, particularly Mala and Sarah Beganskas for their diligent and conscientious proofreading of the final text. If I have missed acknowledging anyone, my sincere apologies. Finally, and most emphatically, I thank the O'Donnell Foundation in Dallas and the Singaporean Agency for Science, Technology, and Research (A*STAR) for essential financial support during the execution of this work.

Timeline

1927–1941	Born, January 13, 1927 in Germiston, South Africa Schooling in Germiston
1942–1951	Attends the University of Witwatersrand; completes 6 years of medical school interrupted by 2 years of training in basic science
1945	Publishes first scientific paper; *S Afr J Med Sci*
1946	B.Sc. (Wits) (Bachelor of Science degree) Kalahari field expedition
1947	B.Sc. (Hons) (Wits) (Bachelor of Science Honors degree) M.Sc. (Wits) (Master of Science degree)
1948	Director of Research, National Union of South African Students (NUSAS)
1950	President, Student Research Council (SRC) Wits
1951	M.B. B.Ch. (Wits) (Bachelor of Medicine and Bachelor of Surgery degree; equivalent to MD degree in the US)
1952	Marries May Covitz
1952–1954	Oxford University (Overseas Scholarship of the Royal Commission for the Exhibition of 1851)
1953	Views the DNA model with Jim Watson & Francis Crick at the Cavendish Laboratories, Cambridge, UK
1954	D.Phil. (Oxford University) (equivalent to Ph.D. degree) Carnegie Corporation Fellow (USA) Meets Seymour Benzer at Cold Spring Harbor Laboratory (CSHL) Begins thinking about the impossibility of overlapping genetic code Begins thinking about the concept of demonstrating colinearity between the gene and its polypeptide Meets Francis Crick and Jim Watson again at Woods Hole Meets George Gamow at Woods Hole Travels with Watson by car across the United States to Caltech

	Temporary appointment at the Virus Laboratory of UC Berkeley (with Gunther Stent)
1954–1956	Returns to South Africa
	Lecturer in Physiology, Witwatersrand University, South Africa
1956	RNA Tie Club piece - On the impossibility of all overlapping triplet codes
	Leaves South Africa to work with Francis Crick at the Cavendish Laboratory, Cambridge, UK
1956–1992	Scientific Staff, Laboratory of Molecular Biology (LMB), Cambridge, UK
1957	On the impossibility of all overlapping triplet codes in information transfer from nucleic acids to proteins published in PNAS
	Seymour Benzer and others spend a sabbatical at the Cavendish Laboratory
1959	Elected Fellow, King's College, Cambridge University
	Electron microscopic work on viruses; dark-field microscopy
	Attends Copenhagen meeting where François Jacob talks about factor X (mRNA)
1960	Pivotal meeting at King's College where the existence of mRNA is postulated
	mRNA experiments with François Jacob and Matt Meselson at Caltech, Pasadena, California
1961	mRNA paper published in *Nature*
	General nature of the genetic code for proteins published (with Francis Crick)
1962	New Laboratory of Molecular Biology (LMB) officially opened by Queen Elizabeth II
	Key discussions with Crick about future research in higher organisms
	La Tranche vacation with François Jacob and families; work on DNA replication with Jacob begins
1963	Begins *C. elegans* work
1964	An unstable intermediate carrying information from genes to ribosomes for protein synthesis published in *Nature* (with François Jacob and Matt Meselson)
	Co-linearity of the gene with the polypeptide chain published in *Nature*

1965	Nonsense triplets identified
	Elected Fellow of the Royal Society
	Elected Honorary Foreign Member, American Academy of Arts and Sciences
1967	Opal suppressors identified
	Honorary D.Sc. degree, Trinity College, Dublin
1968	*C. elegans* work in full swing
1969	John Sulston joins the Brenner laboratory at the LMB
1972	Honorary D.Sc. degree, University of the Witwatersrand, South Africa
1974	First *C. elegans* papers published
	US National Academy of Sciences calls a moratorium on recombinant DNA research
	Appears before Ashby Committee on recombinant DNA
	Bob Horvitz joins the Brenner lab
1975	Ashby report released in the UK
	Attends Asilomar conference on recombinant DNA, Asilomar, California
1976	Honorary D.Sc. degree, University of Chicago
	Francis Crick moves to the Salk Institute, La Jolla, California
1977	Proleptic Director, LMB
	Elected Foreign Associate, US National Academy of Sciences
	Myosin gene from *C. elegans* cloned
1979	Elected Foreign Member, American Philosophical Society
	Elected Fellow, Royal Society of Edinburgh
1979–1986	Director, LMB, Cambridge, UK
1981	Honorary LLD degree, University of Glascow
1981–1985	Non-Resident Fellow, Salk Institute, La Jolla, California
1982	Honorary D.Sc. degree, University of London
1983	Honorary D.Sc. degree, University of Leicester
	Visits Singapore
1985	Honorary D.Sc. degree, University of Oxford
	Honorary Fellow, Exeter College, University of Oxford

1985–1989	Visiting Fellow, Central Research and Development Department, E. I. du Pont de Nemours
1986	Publishes monograph on *C. elegans* nervous system
1986–1991	Director, MRC Molecular Genetics Unit, Cambridge (Addenbrooke's Hospital)
1987	Institute for Molecular and Cellular Biology (ICBM) opens in Singapore
	Companion of Honor
1987–1990	Editor-in-Chief, *J Mol Biol*
1987–1991	Visiting Professor in Medicine, Royal Free Hospital School of Medicine, University of London
1989	Fugu project initiated
1989–1991	Scholar in Residence, The Research Institute of Scripps Clinic, La Jolla, California
1989–1996	Honorary Professor of Genetic Medicine, University of Cambridge Clinical School
1991–1994	Visiting Member, The Research Institute of Scripps Clinic, La Jolla, California
1992	Elected Member, French Academy of Sciences
1992–1996	Associate of the Neurosciences program, New York
1994–2000	Writes monthly columns for *Current Biology*
1995	Honorary D.Litt. degree, National University of Singapore
1996	Honorary D.Sc. degree, Rockefeller University, New York
1996–2001	President and director, Molecular Sciences Institute (MSI), La Jolla/Berkeley, California
1997	Honorary D.Sc. degree, Columbia University, New York
1998–2000	Adjunct Professor, University of California at San Diego, San Diego, California
1999	Honorary D.Sc. degree, University of La Trobe, Melbourne, Australia
2000	National Day Public Service Star, Republic of Singapore
	Complete sequence of fugu genome published
2001	Honorary Doctor of Law degree, Cambridge University
2002	Honorary D.Sc. degree, Harvard University

2003	Honorary Citizen, Republic of Singapore
	Officer, Legion of Honor, France
	Honorary Doctorate, University of Oporto, Portugal
	Order of Mapungubwe, Republic of South Africa
	Honorary D.Sc. degree, Yale University
2004	Founding President Okinawa Institute of Science and Technology, Japan
	Honorary D.Sc. degree University of British Columbia, Canada
	Honorary D.Sc. degree, University of Toronto (bestowed, 2005)
2005	Senior Fellow, The Crick-Jacobs Center for Theoretical and Computational Biology, Salk Institute, La Jolla, California
	Senior Fellow, Janelia Farm Research campus, HHMI
	Honorary Fellow, University College, London
	Honorary Doctor of Law degree, University of Dundee
2007	Honorary D.Sc. degree, Watson School of Biological Sciences, Cold Spring Harbor, New York
	Honorary MD degree, Leeds University, UK

Prizes Awarded to Sydney Brenner*

1968	Warren Triennial Prize
1969	William Bate Hardy Prize, Cambridge Philosophical Society
1970	Gregor Mendel Medal, German Academy of Science
1971	Albert Lasker Medical Research Award
1974	Royal Medal, Royal Society of London
1975	Charles-Leopold Mayer Prize, French Academy, France
1978	Gairdner Foundation International Award, Canada
1980	Krebs Medal, Federation of European Biochemical Societies
1981	Ciba Medal, Biochemical Society
1983	Feldberg Foundation Prize
1985	Neil Hamilton Fairley Medal, Royal College of Physicians
1987	Prix Louis Jeantet de Medecine; Foundation Louis Jeantet de Medecine, Switzerland
	Genetics Society of America Medal
	Harvey Prize, Technion—Israel Institute of Technology Hughlings Jackson Medal, Royal Society of Medicine, London
1988	Waterford Bio-Medical Science Award, The Research Institute of Scripps Clinic, La Jolla, California
1990	Kyoto Prize
1991	Gairdner Foundation International Award, Canada
1992	King Faisal International Prize for Science, King Faisal Foundation, Saudi Arabia
	Bristol-Meyers Squibb Award for Distinguished Achievement in Neuroscience Research, New York
2000	Albert Lasker Award for Special Achievement in Medical Science

*Not a comprehensive listing

2001	Novartis Drew Award in Biomedical Research
2002	Distinguished Service Award of the Miami Nature Biotechnology Winter Symposia
	March of Dimes Prize in Developmental Biology
	Dan David Prize
	Nobel Prize in Physiology or Medicine (with R. Horvitz and J. Sulston)
2003	UCL Prize in Clinical Science, University College, London
2004	Rocovich Gold Medal, Edward Via College of Osteopathic Medicine, Blacksburg, Virginia
	Phillip Tobias Lecture Medal, Wits Medical School, Johannesburg, South Africa
	National Science Award, Singapore
2005	UC San Diego/Merck—Lifetime Achievement Award
	Carter Medal, British Society for Human Genetics
2006	National Science and Technology Medal, Singapore

Prologue

Stockholm, Sweden, December 10, 2002

EVERY DECEMBER 10TH SINCE 1926, the majestic Stockholm Concert Hall resonates with a special air of anticipation and excitement. On this day, commemorating the death of the Swedish chemist, engineer, and innovator Alfred Nobel (most famous for his invention of gunpowder), the King of Sweden presents the annual Nobel Prizes in Physics, Chemistry, Medicine or Physiology, Economics, and Literature. On this evening in 2002, an assembly of Swedish and foreign dignitaries occupy two tiered galleries that span the entire rear width of the massive concert stage (see photo section). The Swedish Royal Family is enthroned on one side of the richly carpeted front stage. Directly opposite them, in the front row of the galleries, are ten immaculately tailored men who are the Nobel Laureates. Quiet descends on the gathered audience, invited guests and all, as Professor Urban Lendahl of the Nobel Committee approaches the lectern to introduce the 2002 Nobel Laureates in Medicine or Physiology.

Your Majesties, Your Royal Highnesses, Ladies and Gentlemen,

We have all begun our lives in a seemingly modest way—as the fertilised egg cell, a tenth of a millimetre in size. From this small cell, the adult human being develops, with its hundred thousand billion cells, through cell division, cell differentiation and by formation of the various organs. To only make new cells is however not sufficient, certain cells must also die at specific time points as a natural part of the growth process. Think for example about how we for a short period during foetal life have web between our fingers and toes, and how this is removed by cell death.

The importance of cell differentiation and organ development was understood by many, but progress was slow. This was largely an effect of our complexity, with the large number of cells and many cell types—the forest could not be seen because of all the trees. Could the task to find the genetic principles be made simpler? Were there a species simpler than humans, but still sufficiently complex to allow for general principles to be deduced?

Sydney Brenner in Cambridge, UK, took on the challenge, and his choice was the nematode *Caenorhabditis elegans*. This may at first seem odd, a spool-shaped approximately 1 millimetre long worm with 959 cells that eats bacteria, but Brenner realised in the early 1960s that it was, what we today would call, "loaded with features". It was genetically amenable and it was transparent, so that every cell division and differentiation could be directly followed in the worm under the microscope. Brenner demonstrated in 1974 that mutations could be introduced into many genes and visualised as distinct changes in organ formation. Through his visionary work, Brenner created an important research tool. The nematode had made [it] into the inner circle of research.

Sydney Brenner, Robert Horvitz and John Sulston. Your discoveries concerning the genetic regulation of organ development and programmed cell death have truly opened new avenues for biological and medical research. On behalf of the Nobel Assembly at Karolinska Institute I wish to convey to you our warmest congratulations and I ask you to step forward to receive the Nobel Prize from the hands of His Majesty the King.[1]

This year's Nobel Prize celebrates the Joy of Worms. Brenner's almost prophetic visions from the early 1960s of the advantages of this model organism have been fulfilled. It has given us new insights into the development of organs and tissues and why specific cells are destined to die. This knowledge has proven valuable, for instance, in understanding how certain viruses and bacteria attack our cells, and how cells die in heart attack and stroke.

The 75-year-old Brenner rises stiffly from his chair with the aid of a cane that has assisted his walking since a motorcycle accident crushed his left leg more than 20 years earlier. Short and stocky, but without overt portliness, Brenner is clad in evening wear replete with starched white collar and swallow-tailed jacket—perhaps the first (and only) time that he has acquiesced to such formality. His famous eyebrows (once described by his long-time friend and scientific colleague, French Nobel Laureate François Jacob, as "enormous, hirsute [and] shaggy")[2] are the defining feature of a ruggedly handsome and expressive face. Brenner's expression is well practiced in registering disdain, boredom, or irreverence. "... gleaming eyes, non-stop animation, midway between leprechaun and gnome,"[3] was how he struck one journalist. Another observed, "Brenner is not beautiful in repose. But the force and intelligence of his features are astonishing. Even as an old man, almost bald, with eyebrows that burst out like the roots of a small forest, he seems to fill the room with vigour. When he starts to talk you are swept along in the icy, buffeting current of ideas, shocked and exhilarated to the point of exhaustion—and still he goes on talking. Profundities, puns, anecdotes and opinions all rush and jumble together."[4]

Brenner is no stranger to scientific accolades. The preceding 37 years witnessed his election to the most august scientific societies in the world, and he has been plied with honorary doctoral degrees and distinguished lectureships from every corner of the globe. Among his repertory of prizes and awards (listed in the preceding section) is the Albert Lasker Award, considered by many the American equivalent of the Nobel Prize. This award was presented twice to Brenner, in 1971 and again in 2000, prompting the observation "the first was for science, the second was for surviving."[5] These and many other relics of distinguished scientific achievement now lie in odd drawers in his many homes, where they were unceremoniously tossed following their receipt. Diplomas for the 20 or so honorary doctoral degrees conferred on him, including those from Oxford, Cambridge, Harvard, and Yale, presumably are scattered about also.

Later that evening at the formal Nobel banquet in the City Hall of Stockholm—an event including about 1300 people and one that many would pay a fortune to attend—Brenner's invited guests are limited to his wife May, two of his four children, and one of his numerous grandchildren. (Nobel Laureates are typically permitted to invite as many as a dozen guests to the festivities. Brenner donated his unused allotment of invitations to Horvitz and Sulston.) As the senior member of the trio, he delivers the customary thank you speech on behalf of himself and his fellow laureates, John Sulston and Robert Horvitz. Typical of all his public addresses, his remarks are laced with irreverence and humor, eliciting amusement rarely witnessed amidst the solemnity of the Nobel proceedings.

Your Majesties, Royal Highnesses, Ladies and Gentlemen,

. . . I was asked to prepare my speech because I understand it will be captioned for the Swedish public, and so it will be difficult for me to change what I have to say, because that will be very confusing. However, I will add something which arises from what we have just seen in the brilliant show that was presented to us, because it gave me many ideas for a uniform for Nobel Laureates. In particular I liked the person who had such a swollen head that [it] could be easily converted into the bubble economy. And for my own choice, if I had any hair at all, I would dye it orange, put on a gold lamé dress and wear a top hat.

But now I come to what I want to say. And the best way I can say it, is to tell you about a letter I've received. A Nobel Prize winner gets many letters. This was from a student in China. His e-mail said: "Dear Dr. Sydney Brenner, I wish also to win a Nobel Prize. Please tell me how to do it." I have been considering the reply, which will say something like this: First you must choose the right place for your work with generous sponsors to support you. Cambridge and the Medical Research Council will do. Then you need to discover the right animal to work on—a worm such as *C. elegans* for example. Next, choose excellent colleagues who are willing to join you in the hard work you will need to do. How about John Sulston and Robert Horvitz for a starter? You must also make sure that they can find other colleagues and students. Everybody will have to work hard. Finally, and most important of all, you must select a Nobel Committee which is enlightened and appreciative and has an excellent chairman with unquestioned discernment.

All of this is necessary to bring you to this point where, on behalf of your colleagues, you can thank everybody for the opportunity to be present here and to make this speech.[6]

PART 1

*Growing Up in
South Africa*

1

A Potent Intellect

That book really turned me on to biology

L IKE THE COUNTLESS EASTERN EUROPEAN JEWS DESPERATE to escape the discrimination and persecution that stained the waning years of the Russian Czarist regime, Sydney Brenner's parents left their Latvian and Lithuanian homes around the turn of the 19th century. Though remotely situated at the tip of the African continent and in so many ways foreign to their cultural origins and tastes, South Africa was touted among Eastern European Jews as a land of financial opportunity and a haven from persecution; for young Russian men it provided an escape from dreaded conscription into the army. This migrant mentality was reinforced by several major shipping lines that actively sought customers from the Jewish communities of Eastern Europe, even employing emigration agents to sell ship passages and advertising widely in local Yiddish and Hebrew newspapers.[1] In particular, the merged England-based Union and Castle shipping lines that provided weekly mail deliveries to the Cape Colony aggressively promoted reasonably priced passage by boat.

For the fortunate among the tens of thousands of participants in this journey, assistance with the financial burden of travel (frequently dictated by unscrupulous shipping agents and corrupt emigration officials) came from relatives already established in South Africa or other favored destinations: the United States, South America, Canada, or Australia. Those less financially secure had to forgo the luxury of choice and allow their futures to be dictated by the most affordable sea passage available. Many settled in Britain, a country considerably less remote from the Continent and one generally hospitable to Jewish refugees of that period. Most knew little if anything of the language and culture of the countries they intended to adopt. Still, they clung to the hope that these far-flung opportunities would offer a life more endurable than that in the Soviet Union.

Reliable information on the size of the South African Jewish population comes from early census reports (the first official census was in 1904). It is estimated that between 1880 and 1910 the initial population of about 4000 swelled to approximately 47,000 by 1911. A further 20,000 Jews are believed to have immigrated during the decades of the 1920s and 1930s. By the mid-1930s Jews comprised about 4.5 percent of the white population of South Africa, a figure that dropped to about 3 percent following the imposition of curbs on Jewish immigration in the late 1930s.[1] Morris Brenner, Sydney's father, left his native Lithuania in 1910 at the age of 19 to join two brothers in South Africa, who in turn had been brought to the country by their father some years previously. Like most Jews enduring the cramped ocean voyage to Cape Town, Morris passed the days and nights in the company of fellow travelers, perhaps participating in hands of *klaverjas*, a popular European card game that relieved the tedium of the long journey and offered distraction from the uncertainties that lay ahead. This wave of migrants also included his mother, Leah Blecher, who left her home in Dvinsk, Latvia, in 1922.

The majority of these immigrants came from rural villages in Eastern Europe, where many had eked out meager livings as petty traders, peddlers, and artisans.[1] Lacking the skills required to farm productively or otherwise live off the land, most settled in urban areas. Indeed, the 1936 census revealed that 96 percent of the South African Jewish population lived in or close to cities and towns.[1] This preference was reinforced by the discovery of diamonds and subsequently gold in South Africa, events that offered prospects for success in virtually every aspect of commercial and, more rarely, professional life.

The many prospectors who sought to mine the rich gold seams buried in the area surrounding Johannesburg included John Jack, a Scot, who came from a farm called Germiston, near Glasgow. When he and a fellow prospector discovered gold near Johannesburg, Jack named the town that sprang up next to his mine after the farm in Scotland. By the early 1920s Germiston boasted the world's largest gold refinery—and a large man-made lake that was to become one of the few distinguishing features of this otherwise unremarkable mining and railroad hub. It was here that the 21-year-old Morris Brenner established a cobbler shop, an enterprise that he maintained for the rest of his life. "My father went on working until he was over eighty. He simply wouldn't give it up, despite the fact that everyone thought that ridiculous. He was one of those people who believed that if you stopped working, you might as well just stop living."

The Jewish community in Germiston was tightly knit, sometimes sponsoring social events to welcome new immigrants. Morris Brenner met his wife Leah at one such function—held in a vacant barn, because the community could not afford the luxury of a more distinguished location. Sydney's sister Phyllis, two years his junior, recalls that their mother, blessed with a delightful singing voice, was often asked to render Russian songs at these gatherings. Morris Brenner, a happy-go-lucky man who cherished music and dancing, apparently fell in love with Leah Blecher the moment he heard her sing.[2]

Brenner's parents were married in 1926 and set up house in two rooms in the rear of their tiny cobbler shop. Sydney was born within a year of their marriage. A modest man totally devoid of intellectual pretensions, Morris Brenner never bothered to become proficient in reading or writing English, seeing little value in these skills for supporting a family. He once (proudly) told a newspaper reporter that he could sign his name, but deferred all documents and correspondence to his wife. Indeed, when Morris was naturalized as a South African citizen 55 years after he arrived in the country, the family engaged a lawyer to complete the necessary documentation. But the elder Brenner became proficient in English and Afrikaans, as well as several African dialects. For many years he enjoyed the daily companionship of an African assistant, Eddie, whom Morris perversely taught to speak Yiddish. Aside from his skills as a cobbler, Sydney's father was, by all accounts, an intelligent and perceptive man with a highly developed sense of humor—distinctive of many Jewish immigrants of that era. "My father was a lot of fun to be around," Phyllis Finn stated. "He was always laughing and joking and loved parties and dancing."[2]

When Morris could afford to move his family to a larger house, he and his wife enhanced the family's earnings from the cobbler shop by renting the master bedroom to boarders. Having worked in a clothing factory after her arrival in South Africa, Brenner's mother Leah helped make ends meet by sewing clothes for her children. "The reality in South Africa was that you weren't really poor unless you were black. But on a relative scale we were poor." Regardless, neither Sydney nor Phyllis remembers having to forgo any essentials. To be sure, the pair eagerly attended Saturday morning movies at the local theater—and could even afford the luxury of a candy or two.[2]

Unlike her agnostic husband, Leah Brenner was sufficiently steeped in Judaic tradition that she maintained a kosher home. Like the majority of Jews of that period, the Brenner family observed the major Jewish holidays and festivals. Furthermore, they revered times when families and friends came together to reminisce about *the old country* and to offer thanks for the stability of their new-found lives. Unconstrained by religious observance, Morris Brenner enjoyed nonkosher cuts of meat, especially rump steaks and jellied pig's feet. But Leah strictly banned nonkosher food from her home, leaving Morris to prepare such offerings on a small gas stove in the garage.[2] Like most of their contemporaries, the Brenners retained their ethnic identity, and Sydney and his siblings were told of the odyssey that their parents had endured in leaving the countries of their birth.

With the outbreak of World War II, Morris enlisted in the South African Army, though a peptic ulcer spared him from joining the many South Africans who fought the Germans in North Africa. Instead, he cobbled shoes and boots at a military garrison in nearby Pretoria, returning home as often as he could.

Sydney Brenner was born in Germiston on January 13, 1927. He retains nostalgic memories of the pervasive smell of leather from his father's cobbler shop, a place that dominated his early life. Together with his sister Phyllis and his considerably younger brother Isaac (referred to as Joe, who being ten years his junior, occupied no special place in Sydney's childhood[a]), Brenner was exposed to the usual duty-bound delights and burdens of an extended Jewish family: maternal and paternal aunts and uncles and numerous cousins who resided in Germiston and in nearby Johannesburg.

Brenner's formidable intellect revealed itself early. At the age of four, he read English fluently. His primary tutor then was an elderly Jewish woman who occupied a room in a nearby house that young Sydney visited with the familiarity of a second home. Noticing his frequent engagement with the newspapers lying about in her home, the woman taught him to read from them. When one of Morris Brenner's customers observed the four-year-old reading in the cobbler shop, she admonished the father for not sending his child to kindergarten. But Morris's modest income could not support the required fees, prompting the concerned customer to arrange for Sydney to attend a facility run by a local church.

Brenner was enrolled in primary school at the customary age of six. South African schooling was closely modeled on the British educational system. Primary school comprised a seven-year curriculum (grades I and II and standards I–V) that preceded five years of high school (forms I–V). His

---

[a] Dr. Isaac "Joe" Brenner currently serves as a senior consulting scientist in geoenvironmental analytical science and technologies. He obtained his Ph.D. in Geochemistry from the Hebrew University, Jerusalem, Israel in 1980. See http://ca.pittcon.org/Technical+Program/tpabstra10. nsf/SCoursesByCat/0BFE2D44784E8CEA852575F20051D7DB?opendocument.

teachers quickly noted his promise, and soon after he entered Germiston Junior (primary) School, he was placed in an advanced class with students two to three years his senior.

Intent on promoting self-education beyond the borders of the United States, the American philanthropist Andrew Carnegie founded a superb public library system in many parts of the English-speaking world. Beginning in the early 1880s the Carnegie Corporation of New York distributed more than $56 million to establish over 2500 free libraries: a fund of $10 million was established specifically for the British dominions of Canada, South Africa, and Australia.[3] Even the small town of Germiston was a beneficiary of Carnegie's gesture, and Sydney eagerly availed himself of the public library's holdings. He devoured books indiscriminately, reading just about everything and anything he could get his hands on. Shortly after Brenner won the Nobel Prize, his mother proudly informed a South African news reporter that her son preferred reading books to traditional boyhood activities. She added that, for as long as she could remember, he spent an absolute fortune on books and noted that when he left South Africa for England in the early 1950s, he packed over twenty crates of books—and a single suitcase of clothes.[4] Brenner's sister Phyllis also retains memories of her brother's passion for books, commenting that he seemed to be "constantly walking around the house holding a book a few inches from his nose, happily oblivious to everything around him."[2]

Among the many volumes that piqued Brenner's boyhood interest was one by the American F. Sherwood Taylor, a prolific writer on chemistry and its history. Published in 1934, *The Young Chemist* provides detailed recipes for aspiring chemists. Sydney was captivated by the book. Resolutely saving his pennies and with help from a local pharmacist, he acquired a few simple chemicals, test tubes, and other bits of basic laboratory equipment. By the time he entered high school in 1936 (at the age of nine), he was executing experiments from Taylor's recipes in the garage of his parents' most recent home. Brenner spent countless hours immersed in this new passion. When neighborhood kids wandered by to entice him to join a game of cricket or soccer, he would impatiently suggest that they invite his sister Phyllis instead.[2]

Keenly appreciative of color from a very early age, Brenner extracted pigments from leaves and flower petals with his primitive chemistry set. As noted in subsequent chapters, pigments and dyes figured prominently in Brenner's early scientific career. An occasional painter, Brenner discovered this talent when confined to bed with a teenage bout of hepatitis. Subsequently, an artist acquaintance encouraged his efforts with watercolors. Both Phyllis and his medical school friend and colleague Phillip Tobias are proud holders of original Brenner watercolors, and several more grace the walls of his home in Ely, England. He even dabbled in sculpture at one time.

The fact that he was two to three years younger than most of his high school classmates in no way thwarted Brenner's booming self-confidence. A former high school classmate recalls that Sydney was frequently in trouble with teachers for inattentiveness or for questioning their pronouncements. When once informed by an irate English teacher that he would surely fail his matriculation examination, Brenner boldly wagered ten shillings (about 25 cents in those days) that he would not only pass the exam, but would additionally earn a first class pass.[b] Not surprisingly, Sydney won his bet.[5] Though considerably shorter than most of his classmates, Brenner escaped the attention of school bullies. But he did not escape the harsh discipline occasionally meted out by teachers. Spare the rod and spoil the child, an attitude adopted from British culture, was standard educational practice in South African schools in the late 1930s. Sarcastic tongue-lashings were frequent and neither students nor their parents considered challenging the occasional canings that occurred at school. Brenner vividly recalls that one of his high school teachers was a stickler for boys wearing ties. If a student attended class without a tie, the teacher would become absolutely irate, sometimes fixing a skylight cord around the neck of a tieless student and forcing the victim to stand on the tips of his toes in front of the entire class. "I believe that my present intense dislike of ties stems from one of these episodes."

When the class was studying *King Lear* as part of the English curriculum, Brenner was instructed to recite a speech by Edmund (the main antagonist in the play) that included the word "bastard" on several occasions. In those days any profane word uttered in public (especially at school) was marvelously titillating to young children. Brenner did so in a deliberately exaggerated manner, prompting his teacher to dismiss him from the class. "The teacher told me that I was behaving like a dog and in order to return to my seat I had to crawl on my hands and knees and remain in that posture next to my desk until I had permission to reclaim my desk chair." Brenner was uninspired by the style of learning at school and applied himself only to the extent required to pass obligatory tests and examinations. Memorizing passages from books and regurgitating information in examinations was not his forte.

---

[b] A first class pass in the South African high school system meant that the student achieved an overall "A" grade for the examination, a level of excellence that essentially guaranteed entry to any university in the country.

"I wasn't the top student as far as grades go. No one pointed to me and said, 'He's going to be a winner.' But I passed all my exams and matriculated with an overall first class pass."

Like most Jewish boys of that era, Brenner took Hebrew lessons several afternoons a week, and at age thirteen he celebrated a traditional bar mitzvah. Phyllis recalls that the local Hebrew school was situated in a section of Germiston with a rough and tumble Afrikaans community that sometimes taunted Jewish youths who ventured into the neighborhood. Noting his uneasiness, his sister once took Sydney by the hand and accompanied him to Hebrew school, but not before she retrieved a knife from her father's cobbler shop that she tucked under her belt to reassure her young brother. "I date my atheism to one such encounter. I was beaten up and sat dazed in the street whispering a Hebrew prayer. But God didn't come. So that was the end of my relationship with him!" But, while rejecting the religious aspects, Brenner enthusiastically absorbed the humor prevalent among immigrant Jewish groups, his father being a particularly influential source. "I suppose that much of my sense of humor came from the gestalt of being a member of a Jewish family. I've always thought that whereas most people's lives consist of drama on the one hand and comedy on the other, for Jews life consists primarily of melodrama and farce. Everything is exaggerated!" Brenner sometimes extended his busy high school (and later university) weekdays with an early morning detour to the local synagogue to join a *minyan* (the quorum of 10 Jewish males over the age of 13 required to formalize religious services), a gesture that facilitated prayers for recently deceased family members of the Jewish community. Not inconsequentially, he received a token fee for these services.

Time permitting (there was precious little of it because of school, Hebrew lessons, and attendance at synagogue), Brenner participated in the usual teen-age activities with school and neighborhood friends, many from the predominantly Afrikaner families that populated Germiston. Here he learned to speak Afrikaans (a language utterly useless outside South Africa) moderately fluently. And contrary to his mother's contention that sports held no interest for him, Brenner played scrum half on his high school rugby team, a position for which his short and stocky frame was ideally suited. Besides frequenting the local cinema for Saturday morning movies, Sydney and Phyllis visited the nearby Germiston Lake, a popular spot for picnics and swimming. As they grew older, each became absorbed in his or her own favored activities. However, Brenner was not above teasing his older sister about her male companions. If he knew that she had been out to a movie on a Saturday night, he would sneak into her bedroom the following morning in search of the box of chocolates that young South African women tradition-ally received from admirers. Brenner enthusiastically approved of his sister's choice of a husband, and, to her delight, he offered a toast to the bride and groom at the wedding ceremony.[2] Much to his mother's horror, Brenner cultivated a beard for this occasion. Eventually the pair struck a deal in which he agreed to wear a suit at his sister's wedding (something he positively hated to do) if he could save his beard.

Brenner's aversion to formal dress surfaced early in his life, a sentiment presumably furthered by his lack of dressy clothes while growing up, but notably reinforced by his disdain for formalities in general. It is rumored that he arrived for his first day of medical school dressed in shorts and sandals. Nor did this attitude appreciably change in later years. Sometime in 1966, he was invited by Sir Lawrence Bragg (the famed English physicist and Director of the Cavendish Laboratories at the time Watson and Crick made their famous discovery) to deliver an after-dinner discourse on the genetic code to a white-tie gathering of The Royal Institution of Great Britain. When Brenner protested in a brief acceptance note: ("I should . . . say that I shall wear a dinner jacket, as I do not like wearing evening dress."),[6] Sir Lawrence replied: "I hope you will not mind the rest of the party wearing white ties, because it is rather late now to tell them to change, but of course it will be quite all right if you came in a black tie."[7] A celebrated photograph of the Governing Board of the Laboratory of Molecular Biology (LMB) taken in 1967 shows LMB Director Max Perutz, Hugh Huxley, John Kendrew, Francis Crick, and Fred Sanger tastefully clad in jackets and ties while an unperturbed Brenner sits tieless and jacketless, sporting a loud shirt with rolled up sleeves (see photo section). Even dressing for the pomp and ceremony of the Nobel Prize in 2002 evoked muted protest. Brenner's late wife May related that when informed that formal evening wear was absolutely required for attendance at the ceremonies, Brenner was of a mind to forgo the entire event.[8]

Brenner's budding interest in science, first awakened by Sherwood Taylor's modest chemistry book, was reinforced by several other texts that he encountered in the Germiston Public Library. Among these was a book by the English novelist, journalist, sociologist, and historian H. G. Wells. Wells attended the London University's Normal School of Science and, inspired by offerings from the famous English biologist Thomas Henry Huxley, he later taught biology. *The Science of Life*, a primer in biology and one of several non-fiction works by Wells, was written in collaboration with his son George Phillip Wells and the British biologist Julian Huxley (a grandson of Thomas

Henry Huxley). Published in 1931, *The Science of Life* found a home in the Germiston Public Library and eventually in Sydney Brenner's bedroom. He was so intrigued by the book that he kept it—unscrupulously informing the library that he had lost it—and paying the obligatory fine. "That book really turned me on to biology."

Close to 1000 pages and extensively illustrated, *The Science of Life* was by no means a children's book. The work was prompted by the commercial success of *The Outline of History-Being a Plain History of Life and Mankind*, a book that Wells published a decade earlier and that at one time sold more copies than the bible in the United States. Intended as a popular work for lay adults with more than a passing interest in biology, *The Science of Life* is divided into nine primary sections that successively address: The Living Body; The Chief Patterns of Life; The Incontrovertible Fact of Evolution; The How and Why of Development and Evolution; The History and Adventures of Life; The Spectacle of Life; Health and Disease; Behaviour, Feeling, and Thought; and Biology of the Human Race. These topics aroused intense excitement in Brenner's fertile mind. In a chapter entitled Vegetable Life, Wells wrote:

> That, in a word, is the secret of the plant. It can drink the pure energy of sun-light, and it can use that energy to build out of the elementary substances that it absorbs, the higher complex molecules of which its tissues consist.[9]

Such passages not only promoted Brenner's fascination with plants, first aroused by his primitive chemistry set, but also gave him pause at the marvel of plants metaphorically drinking the energy of sunlight to build complex molecules out of elementary substances. "How do they do this?" he wondered. The Wells book covered everything then known about biology, and the questions that Wells addressed were exciting to the young science enthusiast. He was fascinated to read that scientists knew how plant pigments participated in photosynthesis. In Brenner's mind, the book literally drew back a veil that enshrouded all of nature. Physics, at least the physics that he had learned in high school, had none of that excitement for him. All he remembers about physics from his schooling is that it was about pendulums. It certainly was not taught as a natural science.

His diligent reading of both *The Young Chemist* and *The Science of Life* instilled in Brenner the fundamental notion that, to acquire new knowledge, books were a good place to start. "I never learned much from attending courses. If I wanted to learn a new subject I would get hold of a book and teach myself. I acquired this habit because there was really no one out there [to whom he cared to listen] to teach me. So out of necessity combined with a natural inclination I learned to rely on books. If one couldn't buy a book one wanted one could almost always find it in a public library—and steal it if necessary!"

As a youth Brenner also cultivated an interest in science fiction, one that he retains to this day. He credits this pursuit to a patronizing uncle who indulged his boyhood passion. This well-to-do relative, one of his mother's brothers, owned a bicycle store in Johannesburg. Much to his delight, whenever Brenner visited his Uncle Harry he would be taken to a nearby hotel for a mixed grill lunch, a rare treat. After lunch his uncle would reach into his pocket and give him half a crown (about the equivalent of an American quarter), a tidy sum of money. Brenner would then run off to a nearby second-hand bookstore that kept old issues of a science fiction magazine called *Amazing Stories*. It was in the pages of this magazine that he encountered the famous names in science fiction—Robert Heinlein, and others. "In later years my knowledge of science fiction stood me in good stead. Francis Crick once wrote a paper on panspermia; the notion that all life comes from outer space. When he showed it to me I scoffed that I'd read about this as a kid." Crick, who, like many others was sometimes taken in by Brenner's mischievous sense of humor, asked Brenner for a reference to the work. "I told him it was in an issue of *Amazing Stories*—in 1936 or 1937, and I added that he should be careful about plagiarism!"

Brenner retained his delight in science fiction, which he found thought provoking because it "provided a different way of looking at the universe and the world—and life." He recalls a famous science fiction story called To Serve Man.[c] In this tale the planet Earth is in a terrible state when it is visited by beings from outer space bearing "humanity boxes" containing endless supplies of energy. Armed with the slogan "to serve man," the aliens offer marvelous courses in which selected earthlings can be taken away and educated. However, no one ever met anybody who returned from one of these courses. The narrator of the story, a journalist, is very suspicious of the motives of these people from outer space. But his friends reassure him, telling him that they are completely altruistic. The story relates the journalist's investigation of what happened to these earthlings, ultimately discovering that the beings from outer space were interested in cooking: *To Serve Man*

---

[c] To Serve Man is a science fiction short story written by Damon Knight that was later adapted for a *Twilight Zone* television episode. See http://en.wikipedia.org/wiki/To_Serve_Man.

was in fact a recipe book! "This kind of duality of meaning always fascinates me. I especially like the play on words and the double entendres in many of them. The wordplay is particularly interesting because it involves a special way of thinking that requires coming up with alternative ideas and explanations. This happens a lot in science—looking at something superficially and then recognizing that there is more than one way of explaining it. It's very much like punning, which I also very much enjoy."

Like many expatriate European Jews, Brenner's parents encouraged intellectual achievement in their offspring as a surrogate for their own limited educational opportunities. Many Jewish South African youths of Brenner's generation with even minimal educational success expected to attend university, preferably to learn a *useful* profession such as medicine or law, professions that offered social status in addition to economic prosperity. Brenner's mother once admonished her elder son never to marry for wealth. But she jokingly hastened to point out that it was just as easy to fall in love with a rich girl as a poor one.

The Brenner family's financial situation made it impossible for them to support a university education for Sydney. However, based on his first class matriculation and his obvious intellectual ability, the Germiston High School successfully nominated Brenner to the local city council for a bursary to attend medical school at the University of the Witwatersrand (in nearby Johannesburg). Brenner and his parents gratefully accepted the opportunity. There was no question in his mind that he wanted to attend university. But he never seriously considered joining the Faculty of Science. In South Africa in those days, that sort of formal training was not an established path to becoming a working scientist. It typically ended with a bachelor degree that might support a career as a schoolteacher. So medical school was really the only viable option for him. "Besides," he scornfully observed in sarcastic reference to the ambition that Jewish mothers traditionally expressed, "my mother really wanted me to become a doctor."

2

In Love with Science

One has to learn chemistry to understand biology

B RENNER ENTERED MEDICAL SCHOOL[a] AT WITWATERSRAND UNIVERSITY in 1942 at the age of 15, three years younger than the majority of his classmates. Early each weekday morning, he would cycle several miles to the Germiston railway station, where he parked his bicycle and boarded one of the commuter trains to Johannesburg. Alighting at the Johannesburg central station, he would walk a mile or so to the main university campus (attendance at the medical school, situated about a mile from the campus, did not begin until the second year of study) to commence a full day of lectures and laboratory exercises. At day's end he would make the reverse trip home. He always brought his lunch from home, having no money to splurge on meals in the student cafeterias.

The University of the Witwatersrand (an Afrikaans term that translates to "ridge of white waters"—a challenge in pronunciation for those not versed in the language) gets its name from a low mountain range that courses through the province of Gauteng. Nearly half of the world's gold supply once came from the extensive Witwatersrand conglomerate reefs first discovered in 1886. Established in 1922 not far from the city center, the university (colloquially referred to as "Wits University," or simply as "Wits") was the successor to the South African School of Mines, a technical college established at the close of the 19th century in Kimberley, the site of a booming diamond mining industry.

Escalating civil unrest in Johannesburg and the surrounding Witwatersrand area forced cancellation of the official inauguration of the university, scheduled for March 1, 1922. The strife was ignited by striking white workers

---

[a] Based on the British model, the South African educational system has no college years (as in the United States) that intervene between high school and graduate school.

in the local gold mines who were incensed by plans to reduce their wages and replace the minority white work force with the abundantly available and considerably cheaper African labor. On March 10, 1922, the recently elected Prime Minister of South Africa, Major General (later Field Marshall) Jan Christian Smuts, arguably South Africa's most charismatic political leader and one of the founders of the League of Nations, proclaimed martial law and directed military action from headquarters that he established in the nearly completed biology block of the new campus. The ensuing four days of violence resulted in hundreds of deaths and injuries. The generally negative reaction to Smuts's handling of this situation cost him and his South African Party the general election in 1924.[b,1]

The university eventually opened in October 1922. Comprising six faculties (Arts, Science, Medicine, Engineering, Law and Commerce), the institution was manned by a mere 73 faculty members who were responsible for meeting the educational needs of a little over 1000 students.[2] By 2004 the student population had swelled to 25,000. Four Nobel laureates (including Brenner) eventually came from its ranks.[c]

Those who wanted to establish a university in Johannesburg hoped to include a medical school. Toward the end of World War I, a first-year curriculum (comprising botany, chemistry, physics, and zoology) was organized at the South African School of Mines and Technology, which had moved from Kimberley to Johannesburg. However, the exigencies of World War I and the immediate post-war period delayed the establishment of a comprehensive medical school curriculum.

The Witwatersrand University School of Medicine was established on the brow of a hill (Hospital Hill) that ascends steeply from the main university campus at Milner Park. Despite the significant distance between the two entities, the choice of Hospital Hill for the new school was a good one. The area was studded with medical facilities (hence the name "Hospital Hill"), including the Johannesburg General Hospital, the Queen Victoria Maternity Hospital, the Fever Hospital, the Florence Nightingale Hospital, the Transvaal Memorial Hospital for Children, and the Non-European Hospital for

---

[b] With the outbreak of World War II, Smuts returned to power as Prime Minister of South Africa and retained this position until his party was defeated by the Nationalist Party in 1948, an era that ushered in apartheid. Smuts died a few years later at the age of 80.

[c] The other Wits Nobel laureates are Aaron Klug, who attended Wits briefly: Nobel Prize in Chemistry in 1982; Nadine Gordimer: Nobel Prize in Literature in 1991; and Nelson Mandela: Nobel Peace Prize in 1993.

non-whites.[d] The South African Institute for Medical Research was also located in the immediate area. In subsequent years, this clutch of health care facilities was further enlarged by the non-white Coronation and Baragwaneth hospitals, facilities that provided much of the extensive clinical experience for which Wits medical graduates soon became internationally recognized.

Though occupied in an unfinished state as early as 1921, the School of Medicine was not completed until the late 1920's. Raymond Arthur Dart, the Australian-born professor of anatomy (who would soon place the medical school firmly on the international academic scene), arrived in Johannesburg in 1923, just a few weeks shy of his thirtieth birthday. A cursory examination of the premises prompted the young professor of anatomy to comment: "Our first inspection of these conditions left my wife, whom I had taken from her medical studies . . . in tears—a woman's prerogative I rather envied at that moment."[3]

The first-year curriculum for medical students was taught at the main university campus in vast lecture halls jammed with several hundred medical and dental students, as well as some from the Faculty of Science. Brenner encountered Edward "Eddie" Roux, professor of botany. A gentle and soft-spoken man with a limping gait, Roux was a knowledgeable and charismatic teacher revered by generations of students. He had acquired a Ph.D. in plant physiology at Cambridge University and taught botany with a modern physiological emphasis, introducing his fresh-out-of-high-school students to such fundamental biological concepts as metabolism, catabolism, enzymology, and biosynthesis—an inspiring contrast to the dull descriptive and taxonomic themes that dominated most British-based botany curricula during the first half of the 20th century. Brenner was taken with this early mentor. "A lot of lecturers used to simply read to us from notes, frequently taken verbatim from textbooks. Eddie Roux's lectures were much more interesting and immediately grabbed my attention."

Roux was once a member of the South African Communist Party and attended the sixth congress of Communist International in Moscow in 1928. He remained politically active in the party until 1935, when he was removed following a purge of alleged right-wing elements in the party.[4] In 1948 (while Brenner was a student at Wits), Roux published a moving account of the

---

[d] During the years of apartheid, hospitals, like many other public facilities in South Africa, were strictly racially segregated.

African struggle in South Africa, entitled *Time Longer Than Rope: A History of the Black Man's Struggle for Freedom in South Africa*. Brenner devoured the book—and was deeply impressed.

Like many of his student contemporaries, Brenner abhorred the politics of apartheid (an Afrikaans word meaning "separateness"), which was enforced by the Nationalist Party following its rise to power in 1948. During this sordid chapter in South African history, the policy of apartheid did not spare higher education. When Brenner was a student at Wits and for many years thereafter, a very modest quota of non-white students (Asians, Indians, and people of mixed race referred to as "Coloreds") was permitted to attend some white universities, including Wits. The tacit understanding was that these students would not mix socially with white students outside of classes. After 1959, when the University Extension Act was promulgated by the government, mixing socially was allowed but only if students first secured authorization from less-than-willing apartheid bureaucrats.[5] Notably, whereas the University Extension Act permitted Indian and Chinese students to attend white medical schools, Africans were excluded, despite a desperate need for doctors for the African population. Furthermore, non-white medical students were barred from contact with patients in white hospitals, although their white classmates enjoyed the privilege of gaining valuable clinical experience in all hospitals. For many years black students were not even allowed to attend autopsy examinations performed on whites. Regardless of many proud declarations that Wits was an "open university," a political observer commented that "it would have been more accurate to label [Wits] University 'less closed' than other South African universities, rather than 'open'."[5]

Student protest against the ruling Nationalist Party was not uncommon, especially in the predominantly English-speaking universities. However, these protests rarely included radical activist groups outside the academic community. Absent structured and determined leadership, they never progressed to frank civil disobedience. Student marches and demonstrations (attended mainly by middle-class and upper middle-class whites) were typically orderly. Having one's name (and possibly one's photograph) documented by the ubiquitous police force, especially the notorious Special Branch, was the only significant risk. Eddie Roux's falling out with the South African Communist Party, possibly reinforced by a general cynicism about the futility of political protest among many liberal white South Africans, led Brenner to a life-long reluctance to join political organizations of any kind. "I have always believed that membership in organizations, especially political ones, compromises one's ideals."

Regardless, Brenner served on the Student Representative Council (SRC), the primary and most influential student organization at Wits University, and in the broader-based National Union of South African Students (NUSAS), organized as a coalition of Student Representative Councils from most universities. In time, NUSAS acquired a politically and socially progressive stance that prompted defection of the Afrikaans-speaking campuses. Ironically, in later years, far more polarized African activists, such as Steve Biko,[e] dismissed NUSAS as too tame for their political imperatives. Irreconcilable differences within NUSAS led to the formation of the rival South African Students Organization (SASO), an entity that was banned by the South African government in 1977.

In addition to political protest, NUSAS (and later SASO) mounted robust educational programs in African communities. Phillip Tobias, Brenner's long-time friend and scientific colleague and now emeritus professor of anatomy at Wits, recalls that when he first became embroiled in South African student politics Brenner was essentially apolitical. After Tobias became president of NUSAS in 1947, he persuaded his friend to serve as Director of Research. This nonpolitical wing of NUSAS organized expeditions to remote areas of the country to promote studies on Africans. Brenner and Tobias toured various universities that belonged to NUSAS, promoting its image—"journeys that were not without danger."[6]

Under Tobias's watchful eye, Brenner rose to the presidency of the Student Representative Council at Wits. His political and social idealism as a 21-year-old university student is best revealed in an impassioned piece entitled Towards a Semantic Sociology,[7] published in the *Student Review-A Journal for Liberals*, a registered student publication. This was followed about a year later by a more dignified treatise called Freedom and the South African Universities.[8] "My thesis in this article," his opening lines read, "is to show that the University stands on three pillars—freedom, knowledge and research—and that these principles are indivisible and alone can bind the University into a unique whole." In a more obvious reference to the insidious

---

[e] Stephen Bantu (Steve) Biko was a celebrated anti-apartheid activist in South Africa in the 1960s and 1970s. As a student leader he founded the Black Consciousness Movement, which later empowered and mobilized much of the urban black population. Continuously harassed by the South African government, Biko died while in prison (ostensibly while being brutally interrogated) and became a martyr of the anti-apartheid movement. His writings and activism sought to empower black people, and he was famous for his slogan "black is beautiful," which he described as meaning: "you are okay as you are; begin to look upon yourself as a human being." See http://en.wikipedia.org/wiki/Steve_Biko.

rise of racial segregation in South Africa, Brenner concluded: "Above all, we must remain free, free to seek experience and to use it. We must say that the only basis for the university is freedom, research and knowledge and that it can only fulfill its important functions if it remains the open society of freed and freeing minds."[8] Such statements almost certainly attracted the attention of the various security arms of the government.

Despite the fact that student protest was relatively innocuous and certainly ineffective, the University of the Witwatersrand was among the more vocal of the English-speaking South African universities. In March 1950 the student newspaper, *Wits-The Witwatersrand Student*, featured the minutes of a meeting of the Student Representative Council, at which Brenner presided. The following formal motions were offered:

1. That a special general meeting consider all the questions arising out of the inter-SRC Conference [recently] held at Durban and that the meeting be asked to endorse once again the principles of academic non-segregation both at Wits and at a national student conference.

2. To reaffirm its support of NUSAS and its principles.

3. Instruct the SRC not to attend any inter-SRC conference if conditions of restriction be laid down as to the race, colour, creed, language, or sex composition of a delegation from Wits.

The article was accompanied by a shoulder-height photo of an appropriately stern-looking Brenner—clad in jacket and tie!

The minutes of that meeting also carried an executive report of a distasteful episode involving Eduardo Mondlane, an African student from Mozambique, whose application for a visa renewal had recently been refused by the South African government.[f] This directive prompted an unusually large protest on the university campus by the SRC, with Brenner presiding. The assembled group passed a resolution expressing their outrage at the government's decision and requesting that it be rescinded. As was usual at protests of this size, uniformed and plain-clothes police and security forces kept a careful watch. Eventually, one of them requested from Phillip Tobias a copy of the

---

[f] Mondlane was refused a visa and was forced to return to Mozambique. Wits University arranged to have lecture notes forwarded to him and even sent someone to Mozambique with a set of examinations so that Mondlane could take them officially. He passed this examination and was awarded a degree by Wits University in absentia. Mondlane became heavily involved in a movement to liberate Mozambique from Portuguese rule and was assassinated by a letter bomb in February 1969. See Tobias PV. 2005. *Into the Past: A Memoir*, p. 62. Wits University Press, Johannesburg, South Africa.

resolution. Seizing on this opportunity to embarrass the authorities, Tobias loudly answered the request: "You'll have to ask the chairman of the meeting," and motioned urgently to Brenner, who confronted the officer, asking him what he wanted. "My minister [referring to the South African Minister of Justice] wants to know what you bastards are getting up to at the university," the man snidely replied.[9]

Brenner immediately brought the incident to the attention of Humphrey Raikes, principal of the University of the Witwatersrand and a man destined to play a significant role in guiding Brenner's academic future. A confirmed liberal and staunch supporter of student political activism, Raikes promptly responded to Brenner's letter.

21st August, 1950

Dear Mr. Brenner,

   . . . I would . . . like to say that I have very greatly appreciated having you as President of the SRC, and also the very fine degree of co-operation which you have displayed.

   I would assure you that I am completely convinced that academic non-segregation, which seems to me to be a dreadful term, is nevertheless the only possibility for fruitful development in higher education in this country. I am quite sure that the African students who have passed through this University have profited by the fact that they have been educated in close association with their white colleagues, and I think too that the white students have profited from their association with the Natives. My policy is however, at all times festina lente, because we must not outrun the constable. We must go slowly ahead on our chosen path and not worry about the consequences. . . . What we have got to do is to do what we think right, and in the latter we shall come out on top.

Yours sincerely,
Humphrey R. Raikes.[10]

Although Brenner was never openly harassed or formally persecuted for his political attitudes and liberal sympathies, the security arm of the governing South African Nationalist Party certainly had his name on one or more lists of political activists. When he applied for a passport to study at Oxford University in the early 1950s, his application was initially denied. When Brenner visited the appropriate office to determine the problem, an official retrieved a weighty dossier that he briefly examined. "This is going to have to wait for a decision from Pretoria [the administrative capitol of South Africa]," the official sternly announced. Brenner was eventually granted a passport, but not before he (perhaps somewhat over zealously) entertained various schemes to covertly leave the country. "In those days one could walk into Swaziland and then go to Mozambique to claim political asylum. Or one could purchase a ticket on a boat to England, hide discreetly until the boat sailed and hope to claim political asylum in Britain." Soon after Brenner arrived in England in early 1952, a former university colleague wrote to him.

> ... just in case you think you are not being watched; ... there appeared a note in The Star [a Johannesburg daily newspaper] expressing consternation on the part of the South African government as to the deleterious effects on the minds on South African students at Oxford, of certain political societies at the university.... .[11]

Immediately prior to Brenner's departure to England in 1951, both the Wits SRC and NUSAS wrote warm letters of thanks, informing him of his election to an honorary lifetime membership. NUSAS made particular mention of his invaluable work in building the research department of the organization during his tenure from July 1947 to July 1949 in his position of National Director of Research and Study.

Unbeknown to most students taking his popular botany course, Eddie Roux maintained an active research laboratory. Brenner eagerly sought permission to visit the facility, filling occasional afternoons observing experiments. Roux and his technician were then purifying various plant pigments by column chromatography. "I was utterly fascinated by this, especially since I had an interest in plant pigments since the time of my chemistry dabbling at home." His busy class schedule precluded formal engagement in research. Still, Brenner was excited to witness a modern biology research laboratory in operation. Chemistry and physics (at least as taught at Wits in the 1940s) were less interesting to Brenner. Nonetheless, the chemistry course was memorable for the fact that he was once ejected from the class for arguing with the lecturer!

During his second year of study, Brenner attended the medical school on Hospital Hill to study anatomy (including histology and embryology) and physiology (which included biochemistry). Guided by Ernest Starling's textbook *Elements of Human Physiology*, Brenner found physiology "totally fascinating," while biochemistry, which in those days focused almost exclusively on intermediary metabolism, opened his eyes to the richness that one can derive from molecular explanations of living processes.

In the anatomy course Brenner encountered an important mentor and role model in Raymond Dart, who was then one of South Africa's most visible (and controversial) scientists. Dart's reputation was based on his groundbreaking contributions to paleontology and physical anthropology. Not long after his arrival in South Africa, one of his students returned from her summer vacation in the Northern Cape flushed with excitement. She presented Dart with a fossilized baboon skull that she had acquired from the Northern Lime Company near Taungs (now called Taung), a small town in the northern Cape Province. Intrigued by this finding, Dart consulted a veteran geologist at Wits, R. B. Young. Young informed Dart that he knew the place well and, coincidentally, was scheduled to visit the Taungs area a few weeks hence. Dart implored him to keep an eye out for further fossils. In due course, two boxes crammed with assorted finds arrived on his desk. In his 1959 memoir, *Adventures with the Missing Link*, Dart wrote:

> As soon as I removed the lid [of the second box] a thrill of excitement shot through me. On the very top of the rock heap was what was undoubtedly an endocranial cast or mold of the interior of the skull. Had it been only the fossilized brain cast of any species of ape it would have ranked as a great discovery, for such a thing had never before been reported. But I knew at a glance that what lay in my hands was no ordinary anthropoidal brain.[12]

Once the fossil specimen was freed from its surrounding breccia, Dart recognized that it represented a hominid of about five years of age. But the so-called Taungs child was neither human nor ape. Specifically, he noted that the fossil molar and premolar teeth, though human in appearance, were larger than the teeth of modern man at the same age. Additionally, the creature was not large brained and large jawed, as would have been expected if the skull were that of an ape. In 1925 Dart published his observations in the prestigious British scientific weekly *Nature*. He entitled his article *Australopithecus africanus: The man-ape of South Africa*.

Dart's contention that the Taungs skull is "logically regarded as a man-like ape"[13] prompted scorn from members of the international anthropological community. For many years he felt deeply wounded by the critical onslaught. He steadfastly refused to donate the fossilized skull to the British Museum and kept it in a safe in his office. With subsequently important anthropological finds in South Africa and elsewhere on the African continent, Dart was duly accorded professional recognition. In retrospect, his great contribution was that he forced the world of paleoanthropology to realize that there had been, at one time, small-brained members of the family of man.[14] Dart's compelling style, his abiding interest in students, and in later years his scientific fame, inspired more than a few medical students to consider joining the ranks of physical anthropologists. As this book will shortly tell, Brenner was among those who came close to seriously opting for such a career. Many years later (in 1982), Brenner penned his former mentor on the occasion of his ninetieth birthday.

Dear Professor Dart,

I am delighted to be able to send you a message of congratulations on the occasion of your 90th birthday. It is a pleasure for me to recall my amateur dabblings in paleontology under your guidance and in the company of Phillip Tobias and other contemporaries.[15]

One of Dart's many legacies to Wits was the farsighted (and far-reaching) establishment of an extra year, following the second year, of medical school, specifically designed to introduce selected medical students to research in the anatomical sciences (including paleoanthropology) and physiology. Despite initial protest from the Faculty of Science, Dart was granted permission to formally organize this extra year of study, which led to a bachelor of science (B.Sc.) degree and secured the return of involved students to medical school when the year was completed. In later years Dart obtained permission for the brightest medical science graduates to complete a second year of research, which led to an honors degree (B.Sc. [with Hons]). Brenner, Tobias, and a number of other distinguished South Africans broke their scientific eyeteeth on these pivotal *science years.*

In addition to Dart, Brenner encountered Joseph "Joe" Gillman, another important mentor who reinforced his growing passion for biology. Like Roux, Gillman was a stimulating (though often intimidating) lecturer with a solid understanding of physiology and biochemistry. He was also a teacher who strived to infuse genuine interest in biological research in his students, knowing full well that with few exceptions (such as Brenner), they were destined for the mere practice of medicine or dentistry. Tall and rangy and possessed with firm and frequently controversial scientific and political opinions that he expressed with a booming Lancashire accent, Gillman's scorn was feared by medical students. Equally intimidating was his partner, Christine Gilbert, a tight-lipped, humorless embryologist whose lectures left students utterly bewildered. Gillman was a powerful, if eccentric, personality. Students found him difficult to get on with, yet he inspired them with his scientific zeal and especially his staunch individualism, traits which greatly appealed to Brenner.

Gillman and Gilbert established a sizable baboon colony on the roof of the medical school. Students conducting research in the Department of Physiology were required to assist in studies on the reproductive cycle of the baboon—not an appealing assignment as it entailed obtaining vaginal smears from female baboons. Christine Gilbert gave but a single demonstration on how to obtain a vaginal smear from a baboon without sustaining a nasty bite. Then one was on one's own! More than a few Wits research students dropped physiology to avoid enduring the baboon colony.

Besides his fearsome reputation among students, Gillman, together with his younger scientist brother Theodore "Teddy," and Christine Gilbert, produced a remarkable series of contributions to many aspects of malnutrition in South Africa. In 1961 Gillman accepted an invitation from President Kwame Nkrumah of Ghana (formerly the British Crown Colony of the Gold Coast) to establish the National Institute of Health and Medical Research in Accra. It is rumored that following several attempts to assassinate Nkrumah, one of his friends and advisors warned Gillman: "[I]f Nkrumah goes, you go too—and you know what it means to go?"[16] Gillman wisely heeded this counsel and retired to England. Not much later Nkrumah's government was overthrown in a military coup. Shortly after arriving in England to study at Oxford University, Brenner and some of his South African friends were reminiscing about Wits. "We joked that one of these days Nkrumah was going to arrive in London pleading for asylum because he couldn't stand being in the same country as Gillman!"

During his second year at medical school, Brenner encountered a volume called *Perspectives in Biochemistry*. This treatise, comprising thirty-one essays edited by the celebrated Cambridge biochemist (and China authority) Joseph Needham and his colleague David E. Green, was a Festschrift presented to the British biochemist and Nobel laureate Frederick Gowland Hopkins by past and present members of his laboratory on the occasion of his 75th birthday. Brenner was intrigued by the volume. "This book opened my eyes to the great richness that could come from the molecular explanation of living processes. I believe it was then, in about 1943, that I realized that one had to learn chemistry to understand biology and that there ought to be a formal discipline devoted to the study of cells at a fundamental molecular level, a discipline that brings life and chemistry together."

These vague notions of what would soon mature to the discipline of molecular biology were enhanced by Brenner's growing interest in genetics. This interest, in turn, was stimulated by his reading an article by the horticulturalist Rose Scott-Moncrieffe entitled "The Biochemistry of Flower Colour Variations," one of the essays in *Perspectives in Biochemistry*. Though initially prompted to read the article because of his curiosity about pigments, Brenner was excited to discover that it addressed the biochemistry and inheritance of pigment production in plants. It concludes with the provocative statement: "These genes, which control series of chemical reactions, are strong though novel weapons with which to attack the problems of biosynthesis." The article revealed to Brenner that much was known about the various plant pigments that he had dabbled with as a schoolboy, and led him to appreciate the notion of biochemical genetics. "This was a kind of beginning of an interest in genetics in my own mind. Of course I knew nothing about genes then because we had absolutely no teaching on this topic."

By this time, in the mid-1940s, biochemistry (recognized as the interface of biology and chemistry) was chiefly concerned with biosynthetic chemical processes in living cells, a field that developed from and largely came to replace what was then called physiological chemistry. The subject then mainly dealt with extracellular chemistry, such as the chemistry of digestion and of body fluids. When interest shifted to cells, it initially focused on the mechanism of the generation of energy and the function of enzymes in cellular metabolism. Intracellular organelles such as the nucleus, mitochondria, the Golgi apparatus, and the endoplasmic reticulum were well-defined morphological entities, but their study was largely restricted to the light microscope. This limitation was sometimes reduced by the application of histochemistry and cytochemistry, specialized fields that incorporated microscopic observations on suitably stained tissues and cells.

In the mid-1940s when Oswald Avery, Colin MacLeod, and Maclyn McCarty published their historic experiments demonstrating that the genetic material in the nucleus was, in fact, DNA, not everyone interested in the molecular basis of inheritance embraced the notion. Despite the rapid progress made by the American geneticist Thomas Hunt Morgan and his collaborators in establishing classical genetics in the fruit fly *Drosophila*, understanding genes and gene function had not fundamentally progressed much beyond the days of Mendel. While young Brenner, impressed by the wonders of biochemistry and physiology, intuitively recognized the potential power of a synthesis of chemistry and biology, the main body of biological knowledge was still firmly rooted in notions and concepts that grew out of the 1930s and early 1940s.

Brenner found precious little to admire in the Wits medical school curriculum—or those who taught it. After reading a chapter in *Perspectives in Biochemistry* on The Disintegration of Haemoglobin in the Animal Body by the noted German-Australian biochemist Max Rudolph Lemberg, Brenner came to the conclusion that the information presented in formal lectures on this topic was hopelessly muddled and out of date. "What amazed me most was that my lecturers had never even heard of Rudolph Lemberg, who was an authority on porphyrins," he lamented. "This was one of the first times that I came to realize that I shouldn't listen to everything I heard in lectures, because my teachers were often uninformed. My growing interest in textbooks and the scientific and medical literature also reinforced my conviction that if one wants to learn a new subject one doesn't need to take courses. It's all there to read and learn about in the literature." His reliance on the scientific literature and textbooks was to serve Brenner well in his future career.

Much of Brenner's second year as a medical student was consumed with learning gross anatomy, a subject then taught at Wits in exquisite detail. There was no shortage of donated or unclaimed cadavers (especially from the impoverished African population). Groups of six medical and dental students were assigned a formalin-fixed cadaver that they were responsible for dissecting, literally from head to toe, over the course of the academic year.

Time not spent in the dissecting hall or in the histology and physiology laboratories was devoted to didactic lectures in anthropology, histology, embryology, and physiology. Brenner enjoyed histology, a subject mainly learned by viewing a comprehensive set of glass slides under the light microscope. He found peering into a microscope and examining individual cells fascinating. All in all, it was an intense year of study: forty weeks, comprising lectures and laboratory exercises that began at 8:00 A.M. each weekday and usually ended well after 5:00 P.M.

During his second year of study, Brenner was startled to learn that if (as fully anticipated) he passed his annual examinations and progressed smoothly through the medical curriculum, he would graduate from medical school at the age of twenty. Since one could not be legally registered as a practicing physician in South Africa until age twenty-one, he was notified that be would have to find something else to occupy that interval—or spend a year in limbo. At that time, Brenner learned of the science year alternative for medical students instituted by Dart in 1926. A number of the faculty in the departments of anatomy and physiology had pursued B.Sc. and B.Sc. (Hons) degrees. Some even had Master degrees in the biological sciences and were actively engaged in research. Brenner unhesitatingly chose to pursue the science year of study.

3

The Science Year Alternative

The Sterkfontein hand—a royal flush in hearts

THE SCIENCE YEAR WAS A FORMATIVE EXPERIENCE. Brenner's class of about a dozen students was exposed to seminars and discussions as well as advanced laboratory exercises in anatomy, neuroanatomy, histology, and embryology—a far cry from the tedium of the medical curriculum. Brenner constructed a model of a chicken embryo using a camera lucida. He carefully traced the outlines of serial sections of the embryo on thin slices of beeswax while viewing sections in the camera and then ordering the slices to generate a large three-dimensional model. He even used this technique to reconstruct a portion of a bush baby brain for a neuroanatomy project. Inspired by the robust tradition of paleontology and physical anthropology in the anatomy department at Wits, he and his classmates embarked on field trips to the famous South African anthropological sites explored by Dart and his colleague Robert Broom. Sometimes joined by Dart himself, they visited the Sterkfontein caves, Makapansgat and Kromdraai.

Brenner was intensely stimulated by his exposure to paleontological and geological fieldwork. The travel that these trips entailed was another much appreciated benefit. Other than a brief sojourn in Cape Town to recover from a bout of jaundice under the care of his father's older brother, Brenner had led a rather insular existence. He was familiar with Johannesburg and its immediate environs, but had seen virtually nothing of the beautiful and bountiful South African countryside. "We'd dig and sort through rubble all day. But at night we'd sit around the campfire playing poker, using sheets of toilet paper as scrip. Everyone knew about the Taungs skull of course. But one night I found what I called 'the Sterkfontein hand'—a royal flush in hearts! I was so excited I woke everyone in the camp at 2:00 A.M. to show them the hand."

These were busy and stimulating times. Brenner might be out in the field and up all night on some weekends. However, on Monday he would be back in the laboratory studying neuroanatomy, and on Tuesday he might be exploring the world of histochemistry. "It was a time when everything and everyone was on the go."

Life was not centered entirely on work, however. Brenner cultivated new friends, many interested in more than simply becoming doctors—or even anthropologists. The majority of this set came from the world of music and the other fine arts at Wits, all well practiced in the sophisticated airs typical of university students. The group engaged in all sorts of entertaining activities, especially skits and cabaret acts, sometimes performed before admiring audiences at the medical school or other suitable venues. These were frequently modeled on a popular English radio program, *The Goon Show*. The gang once contrived a skit on Hector Berlioz's[a] legendary visit to a mortuary. "One of us posed as a corpse covered with a sheet. In due course someone approached the corpse and taking up one of its limp hands, sang: 'Your tiny hand is frozen.' That sort of thing."

Stanley "Spike" Glasser (now a noted musical composer and a Fellow of Goldsmiths College, London University) was the self-appointed leader of this merry band of students brought together by a common interest in university politics and a strong identification with the politically progressive element. The "inner circle" numbered around 10 individuals, but passing associations with other devotees of the avant-garde (including Brenner's future wife May Covitz[b]) sometimes swelled this number.[1] The set relished assuming roles in their plays and skits that called for foreign accents, a talent at which Brenner was (and still is) remarkably adept. In those post-war years mimicking a strong German accent was especially popular.[2] "I think that we were the first real and really loving friends that Sydney had as a young adult," Glasser stated. "It clearly was a vital outlet for him when he was grinding away at medical school or doing research."[1]

---

[a] Like his father, Hector Berlioz was a formally trained physician. His roommate once took him to a public mortuary to purchase an unclaimed body to be taken back to the anatomy dissection laboratory at the university. In his memoirs Berlioz recounted his emotional reactions upon entering this place of the dead in grossly exaggerated terms (for which he was well known).

[b] When May's father arrived in South Africa and was asked for his name he gave his surname, Isaacovitz, the immigration authorities (seemingly arbitrarily) split this into two names and he became Isaac Covitz.

The gang once generated a small movie based on one of Dylan Thomas's stories. Brenner recalls its title: *Portrait of an Artist as a Young Dog.* He wrote the script while Glasser produced the film and the music score. None of the group had ever been to Wales; they hadn't even left South Africa. But this didn't deter them from recreating a Welsh village in Johannesburg from Thomas's descriptions. Perhaps inspired by this experience, in the late 1990s Brenner penned a movie script that he named *How the Quest Was Won,* a spoof on the BBC film version of Jim Watson's famous book, *The Double Helix.* (The BBC film, *Life Story,* starred actor Jeff Goldblum as a gum-chewing Watson.)

During this fun-filled and carefree time in South Africa, Brenner began courting his future wife May, a petite, attractive, articulate, and sophisticated woman, easily able to hold her own against the intellectually formidable and self-confident Brenner. Born into a comfortable upper middle-class Jewish family, May nurtured left-wing political views that matured considerably as a young liberal arts student at Wits University. In 1939 she married Gabriel Balkind, an American-born South African resident described as "an incredibly handsome man who was 99% looks and 1% secrecy."[2]

Soon after the outbreak of World War II, Gabriel returned to the United States with his young wife, where he enlisted in the U.S. Air Force. The couple resided first in New York and subsequently in Los Angeles. Gabriel was away on military duty much of the time, and at one point May lived with her newborn son Jonathan in accommodations rented from a family in nearby Inglewood. Gabriel and May returned to South Africa in December 1946, about six months after Jonathan's birth. To her husband's displeasure May resumed her university studies toward a bachelor degree, majoring in psychology and English literature. Eventually, the couple divorced, and May and young Jonathan moved to a small flat not too far from the university.

May first laid eyes on Brenner at Wits. "Sydney gave my class a single lecture on basic statistics, and I was taken with him immediately," she related. "I had quite simply never encountered anyone like him. He came across to me as a combination of a clown and some sort of Greek god." Being divorced and with a young son, many of May's female friends tried to introduce her to eligible Johannesburg bachelors. But she persistently informed them that the only man she was interested in meeting was a short fellow with ferocious eyebrows who once lectured to her about statistics—but whose name she could not recall. "Sydney was absolutely different from anyone I had ever met. From the very outset we never agreed whole-heartedly on anything.

But we never fully disagreed either. We had the same basic vision of a beautiful future for the world, but we were moving towards this hoped-for future on slightly different paths. He was a medical student and I was studying psychology. We've always been like that. Even nowadays we openly disagree about things and discuss and argue about them. I think that this kept us a more lively couple than we otherwise might have been."[3]

On an occasion when Glasser and his gang were producing one of their infamous skits, May's presence was required for an evening rehearsal (she was a member of the cast). When protesting that she had a small child to look after in the evenings, Brenner gallantly volunteered to baby-sit Jonathan, whom he had not yet met. Upon returning from her rehearsal, May found Brenner and Jonathan deeply engrossed in a game of chess. Recalling this auspicious evening in later years, Jonathan Balkind related that the evening also included the intellectually less challenging game, Snakes and Ladders.[2]

Brenner had few, but equally close, friends at medical school. Seymour Papert, a brilliant young mathematician, and Harold Daitz were among the important people in his life then. Papert and Brenner indulged in many long conversations about science, sometimes into the wee hours. Considered one of Wits University's premier graduates, Papert helped pioneer the discipline of artificial intelligence with Marvin Minsky at MIT in the early 1960s. He was an acknowledged expert on learning theories, known for his studies of the use of technology to enhance learning in children. In the early 1970s Brenner was briefly reunited with his friend at a neurobiology conference at Boston University. Papert suffered serious injuries in late 2006 when he was struck by a motorcycle while attending a conference in Hanoi.

Harold Daitz was another close friend and scientific colleague. "He, and I would have to say Joe Gillman, really taught me the importance of asking questions as a scientist. I credit my life-long pleasure in sitting up half the night discussing science, to Harold. Of course my mother could never understand it when I came home at four in the morning and I told her I that had been talking science. Nor could my wife in later years."

Daitz left South Africa to study at Oxford a year before Brenner. The pair maintained a brisk correspondence. In his first letter to Brenner from London, Daitz offered a poignant description of his initial impressions of the place and how it contrasted with South Africa.

> It is mystical, magical, phantastical . . . it is tame, ordinary, exciting, bracing, inspiring, dreary, drab and sad. In a land like South Africa it is difficult to imagine what this place is like. . . . [The Cape] has the inspiring scenic grandeur . . . superb and unbeatable in its own way; it has the crisp clean

air; it has the winter rain; the old trees and the traces of habitation for 300 hundred years or so. But it has no distinctive tradition . . . [I]t hasn't the population, the background of continuous development from the dawn of the Christian era.

Why do I choose you of all my friends to talk to? Because I feel you may be interested. We shared a good deal of our childhood fantasy as science students and thereafter. We were both misled. We misled ourselves. For we had no guide—no compass. We were dilettantes although we tried desperately hard not to be. There was so much to do—so it seemed. Here one doesn't get quite the same feeling. Here one has little desire to trespass on another field. Or if one must or wants to, you have to do so slowly, carefully, thoroughly. In fact it can no longer be called trespassing. Here one sees achievement.[4]

Brenner's fascination with anthropological fieldwork made him an eager volunteer for formal expeditions. In 1946 he accompanied a group that included several South African geologists to the Kalahari Desert to study its spread and retreat. The expedition generated a scientific publication, Man and the Great Kalahari Desert, on which Brenner was the last author.[c] Several years later he engaged in another field trip to the Kalahari organized by a research team from the University of California. On this occasion Brenner met Alexander Logie du Toit, a prominent South African geologist. Drawn to eccentric and irreverent personalities like moths to a light, Brenner was immediately taken with the elderly du Toit, in whom he recognized a kindred spirit. The geologist had made numerous trips to the Kalahari. But on this occasion he showed up dressed in a black suit with a waistcoat, hat, and tie, looking very much like a Dutch Reformed Church preacher. In contrast the American team was decked out in posh new expedition garb, including brand new bush hats with leopard skin rings around the band.

While traveling along a dirt track in the middle of the Kalahari, the subject of water divining arose. When Du Toit emphatically expressed his belief in this mystical art, one of the Americans looked at him incredulously and asked, "You're a famous geologist and you actually believe in that stuff?" "Of course I do," du Toit blithely replied. "Stop the vehicle and I'll prove it to you." The elderly geologist promptly alighted from the Land Rover and cut a small branch

---

[c] Bosazza VL, Adie R, Brenner S. 1946. Man and the Great Kalahari, *J Natal Univ Coll Sci Soc* **5**: 1–9.

from a tree with evident formality and ritual. He then proceeded to walk solemnly along the road and, lo and behold, the branch began to bend. "Well," said du Toit, "there must be water nearby. Let's find it." The entire party alighted from their vehicle carefully following du Toit, who soon came across a damp patch in the otherwise dry desert, clearly once a small spring. The Americans were astounded. Later in the day, with this demonstration long forgotten, du Toit informed Brenner (in Afrikaans) that he had been in this area as a young man as part of a geological survey and had discovered the spring. He recognized the area, and when the subject of water divining came up, he decided to "put one over on the Yanks."

Thus, life was replete with fieldwork, expeditions, excavations, fossil collecting, and probing the morphogenesis of cave earths. There were also veld frolics to capture elephant shrews in the small town of Bronkhorstspruit and the lizard Agama, northwest of Johannesburg, so that studies could be made of the placentation, chromosomes, and embryogeny of the former, and the brain morphology of the latter. "All conspired to whip up a froth of constructive and mind-expanding activities, not only in the laboratory, but in what American geneticist L. C. Dunn called 'that ultimate laboratory of biology, free nature itself,' wrote Phillip Tobias, who often accompanied Brenner on these sorties."[5]

Tobias continued his career as a paleontologist and physical anthroplogist in South Africa, becoming an internationally celebrated leader in these fields. Brenner, on the other hand, reluctantly decided to forgo further anthropological and geological field work and focus on wet bench biological research. "It was getting to the point where I realized that if I were to continue with that kind of activity I would have to decide whether I wanted to be a professional paleontologist or a laboratory scientist. I chose the latter. It was not an easy decision to make."

In the final analysis, the science year at Wits was a crucial turning point for Brenner. Years later when asked to contribute a commemorative piece to a publication organized by the 1984 science class at Wits medical school, Brenner wrote:

> In those days we were taught an apprenticeship system during which we ... worked through an astonishing array of subjects—microscopy with Dr. Geoge Oettle, tissue culture with Dr. des Lingeris, physical chemistry with Joel Mandelstam, microbiology with Harold Daitz, cytology with Dr. Phyllis Knocker, physical anhropology with L. Wells, anatomy with the master himself [meaning Dart], and a fascinating collection of subjects in what can only be called entrepreneurial biology with the Gillman brothers. We had wonderful lectures from Robert Broom whose pockets

were permanently filled with dinosaur bones and teeth. His influence, together with that of Professor Dart, aroused the interest of a large number of students in paleaontology. I maintained this interest for many years until it threatened to become more than a hobby.[6]

In addition to absorbing the general principles of research and the scientific method, Wits students in the science year were required to successfully negotiate a course in the History and Philosophy of Science. Taught by the Department of Philosophy at Wits, the course was primarily designed for liberal arts students majoring in history or philosophy. So students in the life sciences typically found it somewhat arcane and irrelevant to what they did in the laboratory. Not so Brenner, who was genuinely engaged and stimulated by this unexpected intellectual exposure and who eagerly devoured its content. "I found the course to be extremely interesting. Quite frankly, I hadn't realized that people considered the sciences from such deeply philosophic points of view and from that year on I read a lot on the topic." In 1949 Brenner produced a philosophical essay entitled *Theoretical Biology*, in which he waxed lyrically on "the claims for theory in biology" and went on to discuss how "logic, methodology and systemization can be applied to biological problems."[7]

This largely self-imposed intellectual focus prompted more than a passing interest in contemporary philosophy. It also put Brenner more deeply in touch with the Marxist influences that aspired to improve the lot of Africans in South Africa, and deepened his admiration of people like Botany Professor Eddie Roux. Joe Gillman was of a similar political persuasion, and some of his discussions with Brenner on dialectical materialism prompted the young student to ponder Vladimir Lenin's *Materialism and Empirio-Criticism*. This sort of literary adventure was not particularly unusual. Regardless of the fact that he grew up in a middle-class environment in a small town in a remote part of the world, Brenner used every opportunity to broaden his intellectual horizons and to cultivate the more refined aspects of life. In later years he found the cultural transition from the University of the Witwatersrand to Oxford and Cambridge Universities quite unintimidating.

Brenner was a medical student during the height of World War II, a war that touched the lives of many South Africans, including university students. A good number joined the South African Army and saw action in North Africa and Italy under Field Marshall Smuts. A conscientious home front effort was also mobilized, and during his science year Brenner was required to spend a number of weekends with the 25th Field Ambulance. Dart was the officer in charge of the unit and held the rank of colonel. The medical school group assembled on Saturday afternoons and conducted military training for the duration of the weekend. Brenner was also required to attend occasional military camps lasting as long as two to three weeks, during which time he practiced military maneuvers and brought *injured* soldiers in from the field by ambulance. He rose to the rank of corporal and learned how to drive an ambulance. "I also learnt how to (or should I say how not to) cook from a lecturer in physiology who was in charge of the kitchen!"

The switch from medical school to the Faculty of Science cost Brenner his bursary, which was restricted to financial support as a medical student. But this hardship was largely negated by formal employment as a laboratory technician in the Department of Anatomy. This not only paid his nominal tuition as a student, but additionally provided him a small salary to live on. He was required to assist one of the trained technicians in generating histology slide collections for medical students. In so doing he learned how to process and cut sections of human tissues and how to mount sections on glass slides and stain them with various dyes, another encounter with his childhood passion. Aside from the financial reward of these labors, his employment in one of the medical school's core laboratories reinforced Brenner's interest in histology and the related fields of histochemistry and cytochemistry.

It was as a laboratory technician that Brenner discovered the wonders of a laboratory concoction called Pacini's fluid.[d] "Pacini's fluid has nothing to do with opera!" Brenner cheerfully exclaimed. It was extensively used in the anatomy department to preserve and clear tissues for long-term storage. It had a not displeasing odor, and aware of its high alcohol content, Brenner decided to taste the stuff one Saturday afternoon. The next thing he remembered was waking up later that night lying on the floor of the laboratory and still holding a measuring cylinder in his hand. "I'd obviously had more than a taste of the stuff."

In between full days (and sometimes nights) as a science student and a laboratory technician, interspersed with scientific and military field trips, Brenner's third year at the university passed quickly. He was then faced with another important career decision: to return to his medical school studies—with the foreseeable end-point of becoming an academic physician or, heaven forbid, a practicing doctor—or to pursue another year of research as an honors student. "Most of my classmates went back to medical school. But I decided that I was more interested in science and so I stayed on to do the honors degree."

---

[d] Pacini's fluid was made by dissolving a mixture of water, glycerin (or ethylene glycol), salt, and a corrosive sublimate.

4

Becoming an Independent Researcher

Much to my embarrassment I got it wrong

Honors students in the science curriculum were expected to be appreciably independent and self-reliant in the pursuit of their research goals. Despite his passionate interest in biology, this was Brenner's first serious exposure to bench research, an exposure for which he was armed with little beyond fooling around with a chemistry set in the garage of his parent's home and a year as a science student—this mainly spent fiddling with rather pedestrian histology and anatomy projects. We should recall, too, that he was in South Africa, not exactly a leading light in modern biology, especially in the immediate post-war era. For that matter, nowhere had biology yet achieved the excitement and comprehension that was to explode on the world in the mid-1950s.

As we have seen in the previous chapter, biochemistry was still chiefly concerned with extracellular chemistry. Even when interest shifted to cells, efforts were primarily focused on enzymes and their roles in metabolism. It was known that the nucleus was the seat of heredity and that it contained a mysterious, but to many biochemists a largely uninteresting, polymer called deoxyribonucleic acid (DNA). Despite the fact that Oswald Avery, Colin MacLeod, and Maclyn McCarty had recently published their famous transformation experiments demonstrating that DNA was indeed the genetic material, few outside their immediate circle accepted this notion. It was known, too, that the cytoplasm was teeming with another mysterious nucleic acid polymer called ribonucleic acid (RNA).

Brenner was not yet burdened with the frustrations and failures that attended pressing questions such as the chemical basis of heredity or the mechanism by which the many cellular enzymes and other proteins are synthesized. Immersing himself in the complexities of cellular physiology with his own two hands was reward enough as an honors student, and he became increasingly more convinced that he had identified a career niche. "I decided then that I wanted to be a cell physiologist. I wanted to find out how cells work. This fitted with my early exposures to histochemistry and cytochemistry during my honors year." Long fascinated with dyes and pigments, he began doing cytochemistry to decipher the elements in cells stained by particular dyes, in the belief that such an approach might yield information about cellular functions. He was also excited by the notion that by combining experimental approaches, in which one both examined intact cells under the microscope and also took them apart and ground them up, one might generate some sort of synthesis that would open new vistas in cellular physiology.

Brenner evolved several independent research projects, all of which he managed more or less concurrently with consummate ease. Unfazed by the paucity of scientific equipment (a common situation in post-war South Africa), he improvised as needed. As later pages will reveal, this flair for improvising in the laboratory served Brenner well throughout his career. He attributes much of his histological and cytological prowess to Alf Oettle, a faculty member in the department of anatomy, who, among other things, introduced him to dark field microscopy.[a] Brenner never lost his interest in morphology, and later in his career he used microscopy to great benefit. "I have always believed in what I like to call HAL biology—Have A Look biology. There's something very compelling about being able to see things under the microscope."

Congruent with his fascination with pigments and motivated by Rudolph Lemberg's essay on bile pigments in *Perspectives in Biochemistry*, Brenner learned fluorescence microscopy, then quite a new tool, to examine porphyrins in the liver of Africans suffering from pellagra.[b] This study, initiated and completed early in his honors year, saw the light of day in one of several scientific publications coauthored with his mentors Joe and Teddy Gillman. One of these, "Vitamin A and Porphyrin-Fluorescence in the Livers of Pellagrins, with Special Reference to the Effects of a High Carbohydrate Diet," was published in the *South African Journal of Medical Science*. A second paper,

---

[a] Dark field microscopy is a technique in which contrast between the object and the surrounding field is the reverse of light microscopy, i.e., the background is dark and the object is bright. Certain cellular structures, such as the flagella of bacteria and various parasitic organisms, can be better examined by this technique.

[b] Pellagra is a disease state caused by chronic vitamin B deficiency.

"Porphyrin Fluorescence in the Livers of Pellagrins in Relation to Ultraviolet Light," was published in the British science weekly *Nature*, the first of many contributions by Brenner to this venerable journal.

Another project executed during his honors year prompted Brenner to investigate aspects of the cytochrome oxidase system, an undertaking that required a Warburg manometer. This piece of equipment was not easily acquired, so Brenner built his own, managing to assemble a functional instrument good enough to measure the uptake of oxygen in tissues. These studies introduced him to the Nadi reaction, a cytochemical indicator of the enzyme indophenol oxidase. At that time indophenol oxidase was believed (but not yet proven) to be identical to cyctochrome oxidase, an enzyme discovered by the Russian-born English biochemist David Keilin. Brenner performed a set of experiments that clearly demonstrated a positive Nadi reaction in the mitochondria related to cytochrome oxidase. He wrote a paper on these findings and presented it at the Annual Meeting of the South African Society for the Advancement of Science in 1946.

Brenner encountered further influential scientific texts. Joseph Needham, the editor of *Perspectives in Biochemistry* that so enthralled him several years earlier, published one of these books under the title *Biochemistry and Morphogenesis*.[1] Both Needham's book and a work called *Embryologie Chimique*[2] by the famous Belgian biochemist Jean Brachet, sought to bring a biochemical understanding to embryogenesis. Though often pompous and overbearing, Joe Gillman was a dedicated scholar who quickly recognized that young Brenner was no ordinary student. Motivated to further inspire the promising young biologist, he invited Brenner to join him and his younger brother Teddy at lunch-time sessions during which Brenner was required to read three to four pages of *Biochemistry and Morphogenesis* aloud as a preamble to discussions among the trio. "We would spend hours debating embryological concepts like evocation and induction. It was all rather Talmudic. But Joe was very effective at this sort of Socratic interchange and was a strong formative influence on me in that regard. He wasn't really much of an experimentalist. But he was a fine theoretician."

In the midst of his adventures as an inexperienced cell biologist and avid explorer of the scientific literature, Brenner encountered a text that shifted his focus to genes and genetics more deliberately, a focus that was, henceforth, to occupy a central place in his career. The first edition of *The Cell in Development and Inheritance*[3] by Edmund Beecher Wilson, the renowned Columbia University geneticist, was published in 1896. Inspired by Thomas Hunt Morgan, Wilson developed a theory of sex determination via the X and Y chromosomes and identified the phenomenon of sex-linked inheritance. Brenner was enormously struck by Wilson's views on the chromosomes as the bearers of heredity. "This was the real beginning of my lifelong interest in genes and how they work, and I decided then that I should learn more about chromosomes." But to his frustration he found no one with whom to share these exciting revelations. He tried to discuss genetics with some of his teachers, "but they seemed to understand as little as I did. And my classmates were just a bunch of guys wanting to be doctors." So he set about doing cytogenetics on his own, generating histological preparations of onion root tips that he cut and stained with his own two hands. He gamely battled to absorb Cyril Darlington's *Recent Advances in Cytology*[4] (published in 1932), but found it completely impenetrable. Aided by a book by Darlington and L. F. La Cour, encouragingly called *The Handling of Chromosomes*,[5] Brenner applied his newfound skills to determining the chromosomal content of the South African tree shrew, *Elephantulus myurus*. This substantial undertaking clearly exceeded the expectations for the bachelor of science (with honors) year, and Brenner was encouraged to write a formal thesis and submit it for a master of science degree. "Chromosome Studies in Elephantulus with Special Reference to the Allocyclic Behaviour of the Sex Chromosomes and the Structure of Heterochromatin" was completed a few days shy of Brenner's twentieth birthday.

These and related cytogenetic studies yielded several additions to Brenner's growing publication portfolio. A second paper in *Nature* documented the phenomenon of multipolar meiosis (an occurrence often associated with chromosomal missegregation) in Elephantulus germline cells. But another publication, in the *South African Journal of Medical Science*,[6] caused Brenner some embarrassment. He made squash preparations of testes to identify metaphase plates that could yield the chromosome number. But without realizing it, he counted meiotic cells with the haploid content of chromosomes. Thus, the number he reported was off by a factor of two. "In retrospect I suppose it was an understandable and forgivable mistake." This was Brenner's first experimental foray into the world of genetics and notably stimulated his interest in chromosomes and genes, a field in which he was soon to make his scientific reputation.

Modest though they were, these efforts transpired several years before Francis Crick and Jim Watson produced their model of the structure of DNA—and even before it was conclusively established that *Homo sapiens* had 46, not 48, chromosomes. "Yet," Tobias wrote, "[Brenner] was toying with the nature, biochemistry, and genetics of heterochromatin, as

Prokofyeva-Belgovskaya had been doing in the Soviet Union, . . . [with] Caspersson's absorption spectrophotometry, Bernal and his X-ray diffraction crystallography; [even] Soviet genetics and where Lysenko had gone wrong."[7]

In later years Brenner commented modestly on his precocious scientific breadth. "The circumstances of my scientific education gave me a broad knowledge of biology, and more importantly, there were no boundaries between different subjects. Embryology, neurology, genetics, physiology, and even paleontology were seamlessly joined, something that cannot be said of biology today, where almost everybody is locked into highly specialized compartments."

As a student of medical science, he thought nothing of reconstructing embryos in the morning, performing experiments on the neuromuscular junction in the afternoon, and arguing about philosophy, politics, and life all night. Later, when he began to do serious research, he had acquired enough of the particular form of self-confidence and courage that comes from sheer ignorance to enable him to build ultracentrifuges, synthesize dyes—and even learn enough mathematics—to pursue a life in science.

In 1947, the year that Brenner was engaged in his honors work, the renowned British anatomist, paleontologist, and surgeon Wilfrid Le Gros Clark, then professor of anatomy at Oxford (and later one of a trio of anthropologists who famously exposed the Piltdown fraud) visited South Africa. Like many of his contemporaries and in sharp contrast to notions promoted by Raymond Dart and Robert Broom, Le Gros Clark rejected the notion that the early South African hominids were bipedal. However, Broom had recently described a fossil from the Sterkfontein site that suggested otherwise, and Le Gros Clark wanted to personally examine this specimen. Of course, he also wanted to see the famous Taungs child that Dart kept in a safe in his office.

Brenner was introduced to Le Gros Clark, who, having heard about the up and coming star in South African science, invited him to join his department at Oxford to pursue graduate study. Brenner was, of course, flattered, and for a while was seriously tempted to accept the offer. With the war now over, many South Africans were studying overseas, particularly in London, Oxford, and Cambridge, and Brenner had made a conscious decision to do likewise, though he had no specific timetable in mind. He had no shortage of advice from relatives, friends, and colleagues concerning Le Gros Clark's invitation. Most suggested that he would be wise to complete medical school before going overseas, if for no other reason than for the professional security that a medical degree provided. Dart, then Dean of the

Medical School, gave Brenner similar but more pointed advice. "It sounds to me that you really want to be a biochemist. But most academic positions in biochemistry are in departments located in medical schools. You would do well to finish medical school and then become a scientist if you still want to."

Brenner heeded this advice and returned to medical school. Though he had not as yet evolved a definitive strategy for his professional future, it was increasingly evident to many of Brenner's teachers and colleagues that he was destined for a brilliant career as a scientist, not as a practicing physician. Meanwhile, he was required to return to the oppressive daily grind of lectures, more lectures, and yet more lectures.

5

Failing the Final Year of Medical School

All I could smell was Macleans toothpaste

MEDICAL STUDENTS AT WITS UNIVERSITY returning from the relatively un-structured science year(s) had to painfully readjust to four more years of didactic learning. As with the first two years of medical school, the third year curriculum involved mind-numbing hours of lectures—in pathology, microbiology and parasitology, and pharmacology. The tedium of listening (or not listening) to the professor of pathology and his staff reading verbatim from Robert Muir's *Textbook of Pathology* was occasionally interrupted by the announcement that an autopsy was about to be performed in the hospital morgue. Grateful to be spared another boring lecture, the entire class would troop across the street to the mortuary in the Johannesburg General Hospital, a facility thoughtfully designed and erected as a theater that could seat a hundred students or more to witness autopsies ("post-mortems," as they are called in South Africa). "The year that I returned to medical school was tolerable because we studied pathology and bacteriology and that held my interest to some extent because it was somewhat mechanistically based. But I didn't regularly attend lectures. A friend and I crammed all this stuff in a few months and passed the exams."

In sharp contrast to the heavy emphasis on didactic instruction in the first three years of medical school, the final three years were largely devoted to learning the science (and art) of clinical medicine at the bedside. To this end students organized themselves into teams of five or six. The teams (or "firms" as they are called in the British vernacular) rotated through various teaching hospitals where they received intensive hands-on training in medicine, surgery, the surgical sub-specialties, obstetrics and gynecology, and pediatrics, with comparatively little time in the classroom to accommodate a rudimentary knowledge of psychiatry, forensic medicine, and preventive medicine. For those bent on a career in clinical medicine, this training was nothing short of exceptional. The disadvantaged and economically deprived African, Indian, and Colored populations of South Africa swarmed the health care facilities with textbook examples of acute and chronic diseases. Outpatient clinics were choked by the staggering weight of the sick, and hospital beds for nonwhites were at such a premium that it was the rule rather than the exception for patients to be accommodated on mattresses on the floor placed between beds.

In their fifth and sixth years of training, medical students essentially functioned as interns, even managing busy outpatient clinics in African townships without faculty oversight. This wealth of exposure and direct hands-on involvement spawned generations of highly skilled South African physicians. Attendance at designated clinical rotations was loosely monitored, and during the fourth year of medical school (and sometimes beyond) more hours were devoted to partying and to acquiring expertise in bridge and poker than to learning medicine. Indeed, it is said that more than a few medical students met their pressing financial needs through high stakes poker games that were played long into the night.

No formal evaluation of one's clinical acumen was scheduled until the end of the sixth year of study. Ultimately, this moment of truth had to be faced by three grueling weeks of written examination, oral interrogation, and clinical evaluation of selected patients, some of whom were brought into hospitals specifically for this student ordeal. In essence, during the three years of clinical training, medical students were left largely to their own devices, knowing full well that there would be a final day of reckoning, which, if not successfully negotiated, would require returning to medical school for at least an additional six months—possibly a full year if one's diagnostic skills were seriously deficient.

For Brenner the three years of clinical training were extremely painful. He had absolutely no interest in examining patients or being on the wards, and he found his clinical instructors boring. On a rare occasion when he attended a surgical ward round early in his fourth year, he encountered a surgeon who, in a serious tone, informed the assembled students that surgery was an exact science. At that point Brenner burst out laughing and was summarily dismissed from the class. "I think I must be the only medical student to graduate from Wits without ever laying hands on a patient until the actual final examination."

Brenner blithely devoted his time and energy to his research in the physiology department, firmly convinced that he would be able to finesse the clinical exams—as he had done thus far. But eventually, his lassitude caught up with him. During the examination in clinical medicine, he was instructed to smell a patient's breath and report a diagnosis. The patient was a severe diabetic reeking of exhaled acetone, which any medical student with even a modicum of clinical experience was expected to immediately recognize. Brenner has a very keen sense of smell. Once in the emergency room at one of the African hospitals (another rare occasion of Brenner setting foot in a hospital as a medical student), he correctly diagnosed acetylene poisoning from the smell of an African patient's breath. Africans frequently added calcium carbide (retrieved from gold miners' lamps) to their home brew, a vicious concoction called *skokiaan*.[a] Brenner astutely recalled (perhaps from his boyhood dabbling in chemistry) that acetylene is emitted when calcium carbide is dissolved in water. But when during his final year examination he was asked what he had smelled on the diabetic patient's breath, Brenner confidently replied: "Macleans toothpaste."[b] "I wasn't being facetious," he lamented. "That was really all I could smell. But the examiner failed me on the spot. I did fine with the case previous to that because I got to the patient before the examiner and asked him what was wrong with him. He pointed to his heart and said: 'The doctors always listen here.' So I listened and heard a murmur and correctly diagnosed mitral stenosis."

Raymond Hoffenberg, a native South African and former president of the Royal College of Physicians, recounted that Dr. Frank Forman, professor of medicine at the University of Cape Town and Brenner's external examiner, told him, "I had no choice—Brenner was quite hopeless at clinical examination."[1] Others were equally pained by the overwhelming imperative to fail the obviously brilliant student. The Dean of the School of Medicine, Guy Elliot, wrote to a colleague:

> I am afraid young Brenner came to grief in his final clinical examination and has to repeat the subjects of Medicine and Surgery. One tried so hard to persuade him to adapt his very logical mind to the inconsistencies of clinical work, and something just prevented him from following the advice it seems.

---

[a] *Skokiaan* is a term used to describe a potent, quickly maturing and easily hidden illicit brew that is blended from all sorts of noxious ingredients, including methylated spirits and calcium carbide. (Adam Kuper, *Anthropology and Anthropologists* [Routhledge, 1983], p. 32.)

[b] Macleans toothpaste was (and still is) manufactured by a British company and was a popular brand in South Africa.

The assessment of his clinical ability in the examination I left largely to my external examiner colleagues and things just could not go right.[2]

Brenner's examination in surgery also seemed headed for disaster. The examiner showed him a patient with an exposed leg and asked, "What do you see?" Brenner replied, "A leg." He then asked the student, "What kind of leg?" and Brenner replied, "The right leg!" However, he did pass surgery by the skin of his teeth.

A particularly arduous ordeal imposed on medical students was a six-week sojourn in one of several maternity hospitals scattered around Johannesburg, some as far as the coastal city of Durban, Natal, some four hundred miles away. Students were required to remain on the maternity home/hospital premises at all times and were assigned patients in the early stages of labor. Even though a woman in early labor might not deliver for many hours, especially with a first pregnancy, students were required to monitor the patient's vital signs (and those of her fetus) every few hours, no matter the time of day or night—and more frequently as the second stage of labor approached. When, at long last, the time for delivery arrived, students performed the procedure under the eagle eye of a trained midwife. In nonwhite maternity hospitals students were also required to suture episiotomy[c] incisions at any time of the day or night. (But they were not considered expert enough to perform this function on white patients.)

Medical students lobbied to be assigned to white maternity hospitals where the patient load was considerably smaller than in the nonwhite maternity wards and where one was relieved of the onerous burden of closing episiotomy incisions in the wee hours. Indicative of his general disdain for clinical medicine, Brenner neglected the important pragmatic consideration of identifying a maternity home of his choice. Thus, he was summarily dispatched (together with a few other derelicts) to an African maternity home near the Durban shipping docks.

"It was a hospital at the very bottom of everyone's preferred list. There were about four other medical students there with me and we were housed in a dilapidated hotel where, stating it mildly, amoebic dysentery was probably the least severe illness you could come down with. A constant stream of ladies-of-the-night who plied their trade on the Durban docks occupied the rest of the hotel." Brenner was stuck in Durban for six weeks with literally

---

[c] An episiotomy is a common obstetrical procedure in which incision of the vaginal canal is deliberately used to forestall vaginal tearing during childbirth.

nothing to do except deliver babies—and chat with the prostitutes. Out of sheer boredom, he read the obstetrics and gynecology textbooks and, to his everlasting amusement, achieved the top score in the obstetrics and gynecology written examination.

This unexpected triumph aside, having failed clinical medicine and marginally passed clinical surgery, Brenner had no choice but to extend his training for six months if he wanted to reap the traditional benefit of his six years as a medical student. Adamant as he was about pursuing an academic career in research, he seriously considered forgoing the extra months of study. In the end he heeded the advice of everyone he consulted (including Wits University principal Humphrey Raikes) to finish what he had started.

During his six-month penance, Brenner wisely enlisted a knowledgeable classmate to show him around the wards. When he finally saw his name on the list of students who had completed all medical school requirements ("qualified" in British parlance) he thought to himself, "This is it for me as far as clinical medicine is concerned." In July 1951, a few weeks after he finally finished medical school, Brenner tersely communicated in writing with the South African Medical and Dental Council: "I wish to inform you that I do not intend to register as an intern as it is not my intention to practice medicine."[3]

His disappointing experience as a senior medical student had little, if any, effect on Brenner's expanding self-confidence. While his fellow medical students were scrambling to identify heart murmurs, remove inflamed appendices, and deliver screaming babies, he continued his research. Fortified with modest funds for supplies and small items of equipment by way of several local scholarships and grants, and essentially unassisted and unsupervised, Brenner extended his interest in dyes as probes of cellular function by employing supravital staining, a process in which fresh tissues are stained without the use of fixatives.

In the late 1940s the pioneering efforts of Keith Porter and Albert Claude at Rockefeller University were beginning to yield primitive electron microscopic images of cells. Additionally, though the light microscope remained the primary instrument of investigation for classical cell biologists, its use was increasingly augmented with histochemical and cytochemical techniques, such as supravital staining, prompting the celebrated Belgian biologist Jean Brachet to later write, "Cytochemistry had brought us to the very heart of what much later became molecular biology."[4]

Brenner was keenly aware of Brachet's contributions. He was also intrigued by reports in the literature that whereas DNA was associated with chromosomes, RNA was in the cytoplasm and was associated with particulate entities then called chromidial bodies, or microsomes. By using ultracentrifugation of small pieces of rat liver and then fixing and staining these pieces, Albert Claude (who won the 1974 Nobel Prize in Physiology or Medicine with Christian de Duve and George Palade) had demonstrated that RNA, which readily stained with a supravital dye called methyl green pyronin, was associated with both microsomes and mitochondria. However, he suggested that the staining of microsomes (chromidial bodies) may result from the diffusion of RNA from mitochondria, perhaps because these were injured during the preparation of tissues.

Brenner set about repeating Claude's experiments using a Beams centrifuge driven by compressed air, which he constructed largely on his own. He examined subcellular fractions of liver cells under the microscope and concluded that the chromidial bodies contained both RNA and protein and were clearly separated from mitochondria. Brenner (and others who referred to chromidial bodies or chromidial substance) was, in fact, describing ribosomes, the cytoplasmic seat of protein synthesis. And, of course, the RNA that he and others identified in particulate cytoplasmic fractions was ribosomal RNA. Brenner was proud of these experiments that were ultimately published in the *South African Journal of Medical Science* under the imposing title "The Identity of the Microsomal Lipoprotein-Ribonucleic Acid Complexes with Cytologically Observable Chromidial Substance (Cytoplasmic Ribonucleoprotein) in the Hepatic Cell." This step was beyond pure histochemistry, and he considered the experiments his first direct connection with modern molecular biology. Some years later when he first visited the United States, he was surprised to discover how many people had actually read that paper. Indeed, in early 1948, shortly after this paper appeared in print, Brenner received a letter from the distinguished Scottish nucleic acid biochemist James Norman Davidson, author of *The Biochemistry of Nucleic Acids*, one of the first modern texts on this topic. The letter critically addressed what Davidson considered to be Brenner's incomplete referencing of the literature on nucleic acids, notably his own contributions. Undaunted by a personal communication from this venerable senior scientist, Brenner responded:

17th May 1948

Dear Professor Davidson,

Thank you for your letter of the 23rd March. I am in full agreement with you when you say that our knowledge of the nucleic acids has advanced considerably since 1944: but when my paper was written the position and significance of the nucleic acids in the cell was still unclarified. My main task in that paper was to reveal how little we did know at that time and to attempt to suggest some solution of the problem of nucleic acid distribution in the cell.

In quoting from your paper (1944), I wished to lend authority to the general opinion held at the time that, although nucleic acid was present in all the particles isolated from the cell, the relation between one or other of these particles and the cytoplasmic ribonucleic acid (or chromidial substance) was not precisely known. I apologise if I have misrepresented your views.[5]

Now a member of the international community of cell biologists, Brenner maintained a robust correspondence with scientists around the world, accumulating an impressive collection of reprints. In those pre-Xerox and pre-internet days, collecting reprints was often the only way of keeping up with the literature, especially in South Africa, where scientific journals took months to arrive and where only a fraction of the world scientific literature was subscribed to by university libraries.

It was now 1949 and Brenner had just turned twenty-three. But he was already identified as something special in the South African academic scene, someone evidently cut from a cloth different from even the brightest students that regularly matured from Wits medical school. In truth none of his mentors had previously encountered the single-minded passion for science so evident in Brenner or had his capacity for independence and self-reliance in the pursuit of this passion. There were few problems in the laboratory that he could not troubleshoot himself, including jury-rigging functional scientific instruments as his needs dictated. He had not yet offered any spectacular contributions to biomedical science. Nonetheless, his mentors were convinced that he would attain such distinction once he joined cutting edge laboratories outside South Africa, as most anticipated he would in due course. In recognition of his achievements and perhaps in anticipation of a future of unqualified academic distinction, Brenner was elected to Membership (not to be confused with the more prestigious Fellowship) of the Royal Society of South Africa.

With his self-described dual career as a reluctant medical student and practicing scientist drawing to a close, Brenner began contemplating his future seriously. In 1948 he had been introduced to another distinguished visitor to South Africa, the British embryologist, geneticist, and philosopher of science, Conrad Hall Waddington—affectionately referred to as "Wad" by those who knew him well. Brenner and a fellow student and friend,

George Clayton, were assigned the responsibility of escorting Waddington to and from his many appointments and generally attending to his everyday needs (some on which Brenner declined to elaborate).

Brenner and Waddington became well acquainted and remained friends and correspondents until Waddington's death in 1975. This correspondence included frequent letters imploring Brenner to send him culinary delicacies from South Africa, reflecting the acute shortage of such goodies in post-war Britain. On one occasion he requested that Brenner ship him "tinned meats (bacon, ham, corned beef, Vienna sausages, etc.) with some dried raisins, and candied fruit, particularly figs and pineapple."[6] A few weeks later, he informed Brenner that he would really prefer guava jelly, pineapple and Cape gooseberry jam, bacon, dried raisins, and apricots. Waddington had recently assumed the chair in genetics at the University of Edinburgh. He enthusiastically invited Brenner to join him there. "But he also told me that if I wanted to train at a place with really good biochemistry I should go to Cambridge and work in Albert Chibnall's biochemistry department."

The Department of Biochemistry at Cambridge University was then world renowned. It included J. B. S. Haldane and Joseph Needham, as well as a graduate student named Fred Sanger, with whom Brenner would later enjoy an intimate professional relationship and who would in due course be awarded two Nobel Prizes for his extraordinary contributions to molecular biology. Brenner wrote to Chibnall, the Sir William Dunn professor in the Department of Biochemistry at Cambridge and a leading figure in British bio-chemistry. However, Brenner's association with Cambridge was postponed, and he had to wait a decade before embarking on his scientific explorations there. Who knows how the history of science might have unfolded had Bren-ner attended Cambridge—where Watson and Crick were shortly to embark on their historic quest.

As we saw, Brenner had more than a passing association with the princi-pal of the University of the Witwatersrand, Humphrey Raikes. Raikes had studied chemistry at Oxford, and early in 1951 he suggested that Brenner con-sider working with Sir Cyril Hinshelwood, professor of chemistry at Oxford, who Raikes knew from his days as a Fellow of Trinity College. Hinshelwood, a thermodynamicist and chemical kineticist (who shared the 1956 Nobel Prize in chemistry with the Russian chemist Nikolay Semenov), had recently auth-ored a book called *Chemical Kinetics of the Bacterial Cell*, which Brenner immediately purchased and read. "I never understood a word of it. But this opportunity sounded very promising to me because it seemed to me that Hinshelwood was trying to merge chemistry and biology, something that

I had long believed to be important." Brenner wrote to Raikes expressing his interest in studying under the celebrated English chemist.

> I am deeply interested in the physical-chemical approach to biology and wish to obtain training in physical chemistry, especially as it is applied to biological problems. In recent years, Professor Hinshelwood's main research work has been directed at problems of the chemical kinetics of bacterial metabolism, and it is for this reason that I think that I might be acceptable to him as a student in his department.[7]

Brenner did not know exactly what research he wanted to pursue. Nonetheless, he had resolved to work in an environment in which modern chemistry could be applied to the understanding of cell function, especially the function of genes. He had read Erwin Schrödinger's *What is Life?*[d] shortly after it was published. Many famous scientists who read this book stated that, were it not for Schrödinger, they would have ended up playing the violin for handouts in the London tube stations. But Brenner did not derive any particularly inspirational insights about the gene from this book.

In September 1951, Raikes informed Brenner that he had contacted Hinshelwood, who quickly responded affirmatively. In thanking Hinshelwood for his support, Raikes informed his British colleague that "I do not think that you will be disappointed with Brenner's performance, though, as I have already warned you, he is not yet a good physical chemist."[8] All that now stood between Brenner and a place as a graduate student at Oxford was the obligatory financial support.

On Raikes's recommendation Brenner applied for a coveted scholarship from the Royal Commissions for the Exhibition of 1851, an organization established in 1850 by Queen Victoria to mastermind the Great Exhibition of the Works of Industry of All Nations at the Crystal Palace in London. Consonant with the intention of stimulating industrial education and extending the influence of science and art on industry, the foresighted commissioners also established fellowships and scholarships for science and engineering graduates. Limited to candidates from the United Kingdom and its colonies, these scholarships were highly competitive.

---

[d] *What is Life?* is based on a series of public lectures and was published in 1944. In this book the renowned theoretical physicist Erwin Schrödinger speculated about the nature of the gene and its function as a coding blueprint. This small book purportedly influenced a generation of biologists interested in genes, including James Watson and Francis Crick.

In April 1952 Brenner was informed that he had won a Royal Commissions scholarship. He wrote Raikes a final letter of thanks.

> To be able to go to Oxford University to pursue my studies is for me a great privilege, and I feel sure that were it not for your interest and assistance I would not have accomplished this.[9]

Raikes graciously replied:

> ... [A]s I have always told you, the university tries to help those who help themselves. And if you have not helped yourself I don't know who has.[10]

A congratulatory note from Sir Cyril Hinshelwood followed. In his letter Hinshelwood addressed areas that Brenner might consider for graduate study.

> ... [W]e have not yet had an opportunity of discussing the exact lines of your work. [But] if opportunity offers, you might find it useful and interesting to read either in whole or in part some of the following ... Dubos: *The Bacterial Cell*, Work and Work: *The Basis of Chemotherapy*, D'Arcy Thompson: *Growth and Form*. We have a certain amount of work on yeast going on and if you would like to take a general interest in that Lindegren's *The Yeast Cell* is worth having a look at.[11]

A concluding sentence from Hinshelwood, added almost as an afterthought, particularly heightened Brenner's interest.

> One matter which I did wonder if you would be interested in, though I am not pressing it, is some work on phage. If so, any general information on that subject which you can collect as a background would be extremely valuable.[11]

Brenner enthusiastically responded:

> I shall read the books you have recommended and shall also obtain information on phage, in which I am very interested.[12]

Reviewing Brenner's early exposure to research in the years that culminated in his master's degree in biology and in the succeeding years as a reluctant medical student, one can trace a clear progression of his interest in cells in general to subcellular organelles, finally settling on chromosomes and genes. Almost entirely self-taught and self-motivated, he had arrived at fundamental and contemporary questions in biology: What are genes? How do they work? This was to become the primary focus of his scientific career. His timing was impeccable. At nearby Cambridge University, James Dewey Watson and Francis Compton Crick had already begun their quest to decipher the structure of the gene.

Simply stated, the genetic "coding problem" was the following. Assuming that genetic information did, indeed, reside in the DNA that populated the nuclei of living cells (a notion not at all widely held then), how was this information encoded in genes and how was this code deciphered to yield proteins, the workhorses of the cell, especially the enzymes that catalyze so many life-giving processes? A paper by the celebrated English physicist William T. Astbury, which Brenner spotted in the proceedings of a symposium in Cambridge in 1946, further awakened his interest in the coding problem. Astbury had furnished the first X-ray diffraction patterns of DNA as early as the late 1930s,[e] noting that the DNA structure repeated every 2.7 nm and that the bases lay flat, stacked 0.34 nm apart. At the 1946 Cambridge symposium he noted:

> Biosynthesis is supremely a question of fitting molecules or parts of molecules against another, and one of the great biological developments of our time is the realisation that probably the most fundamental interaction of all is that between the proteins and the nucleic acids.[13]

Astbury also suggested that the similar spacing between the nucleotides and the spacing of amino acids in proteins "was not an arithmetical accident."[13] This similarity in unit spacing was, in fact, fortuitous. But the suggestion of a functional relationship between DNA and proteins resonated with Brenner.

Prior to his departure for Oxford in late 1952, Brenner encountered other contemporary papers that wrestled with the coding problem. One of them, published in 1950, was by Cyril Hinshelwood, who wrote:

> In the synthesis of protein, the nucleic acid, by a process analogous to crystallization, guides the order by which the various amino acids are laid down. . . .[14]

---

[e] During the pre-war years William Astbury at Leeds University began using X-rays to determine the structure of macromolecules in crystals. Astbury took hundreds of X-ray diffraction pictures of fibers prepared from DNA, from which he tried building a model. Astbury concluded—correctly—that the bases lay flat, stacked on each other like a pile of pennies spaced 3.4 Å apart. But he made serious errors, his work was tentative, and he had no clear idea of the way forward. See http://www.pbs.org/wgbh/nova/photo51/before.html.

Brenner also read a paper by Alexander Dounce, who in 1952 suggested that polypeptides were laid down on RNA molecules and that "the RNA sequence determines on a one-to-one basis the sequence of amino acids in the polypeptide."[15] These evocative reports prompted Brenner to evolve his own notion of protein synthesis. High energy phosphate bonds were very much in vogue then, and so he entertained the notion that nucleic acids and proteins were co-synthesized through aminoacyl nucleotide intermediates and that was how DNA specified amino acid sequences. Regardless of the fact that this hypothesis was hopelessly wrong, Brenner's professional future was now seriously engaged with understanding how genes work.

His early interactions in South Africa with his friend Seymour Papert, a budding expert in computation in general, had awakened Brenner's interest in information storage and retrieval. Thus, Brenner was aware of John von Neumann's analogies between computers and the brain and diligently read von Neumann on computers and information storage. He acquired a copy of the *Hixon Symposium on Cerebral Mechanisms in Behavior*, a symposium held at Caltech, in Pasadena, California in 1948. These proceedings included a famous paper by von Neumann[f] in which he introduced ideas on self-reproducing machines, which he called automata. Brenner was especially struck by a passage in von Neumann's paper in which he mentioned that, in order for a self-reproducing machine to work, one needs not only a mechanism of copying the machine, but also a mechanism of copying the information that specifies the machine. "I wasn't smart enough then to realize that what von Neumann was talking about was essentially how DNA might both duplicate itself and specify a code for protein synthesis. But if one thinks about this notion advanced by von Neumann in a historical context, one might suggest that his ideas preceded those of Watson and Crick—except that he didn't know about DNA of course."

On August 27, 1952, friends, colleagues, and former mentors in the departments of anatomy and physiology at Wits University hosted Brenner to a rowdy and emotional farewell. A week later he set sail for England. He had few regrets. He later told Horace Judson, author of *The Eighth Day of Creation: Makers of the Revolution in Biology*: "South Africa was a very

---

[f] John von Neumann was a mathematician who made major contributions to numerous fields, including the development of game theory and the concepts of cellular automata. Along with Edward Teller and Stanislaw Ulam, von Neumann worked out key steps in the nuclear physics involved in thermonuclear reactions and the hydrogen bomb. See http://en.wikipedia.org/wiki/John_von_Neumann.

underdeveloped country scientifically—a provincial country. Facilities for research were really quite primitive. I mean, if you wanted to stain something you had to first synthesize the damn dyes yourself."[16]

Harold Daitz, Brenner's close friend and intellectual collaborator, had also left South Africa to study at Oxford and Brenner was eagerly anticipating the resumption of their relationship. But he had no immediate plans of renewing his relationship with May Covitz, who had journeyed to London with her son Jonathan to pursue graduate studies in psychology. May and Sydney had corresponded, but did not commit to a permanent relationship at that time. "'It's been wonderful knowing you and we should get together in London when we are both there,' we used to say,"[17] May related.

On August 28, 1952, Brenner boarded the Union Castle steamship *Arundel Castle* bound for England. The Sydney Brenner who docked at Southampton on a cold October morning was no longer the youth excited about the mysteries and challenges of biology. He was 26 years old, with a master of science and a doctor of medicine degrees, bursting with ideas and notions of cell function and genetic coding. He was eager to take the next major step in his career, acquiring a doctorate from Oxford and testing his intellectual mettle on the international stage of science. "Otherwise you can't know whether you're really any good or not."

PART 2

The Postgraduate Years

6

Viewing the DNA Model

I knew what I wanted to do with my life

S HORTLY AFTER WORLD WAR II BRITAIN TURNED HER ATTENTION to increasing scientific manpower in the country. A committee that included novelist and science advocate C. P. Snow grappled with this issue and recommended doubling the annual output of active scientists.[1] The government welcomed these recommendations, and new funds were dispensed to British universities. Oxford and Cambridge expressed initial reservations about losing their much-revered independence. Their reticence quickly succumbed to the lure of government funding, and, in due course, Oxford embarked on an ambitious building program.[2] This investment paid handsome dividends. In the 30-year period between 1939–1968, the science departments at Oxford accumulated seven Nobel Prizes.[a]

As South Africa was a colony of the British crown, many South Africans born and raised in the preapartheid era likened visiting the United Kingdom for the first time to meeting a revered grandparent. Brenner's excitement of ultimately setting foot finally in Great Britain, however, was quickly dampened by the shocking news that his good friend Harold Daitz had died shortly before his arrival. "After arriving in London I went directly to his flat, eager to see him again and was stunned to hear from his wife that he had died suddenly and unexpectedly the week before. This was a terrible blow. Aside from the fact that I was extremely fond of Harold, I had all sorts of grand ideas about us doing science together at Oxford."

Brenner paid the required university matriculation fee (a modest six pounds), a college entrance fee of a further 15 pounds, and a tuition fee of

---

[a] Ernst Boris Chain and Walter Florey in physiology or medicine in 1945, Robert Robinson in chemistry in 1947, Cyril Hinshelwood (Brenner's mentor) in chemistry in 1956, Dorothy Hodgkin in chemistry in 1964, Rodney Porter in physiology or medicine in 1972, and Nikolaas Tinbergen in physiology or medicine in 1973.

20 pounds per quarter from his fellowship of 450 pounds a year. He located suitable digs conveniently close to the main university campus where he was quickly introduced to food rationing. This practice, instituted in England during the war, was stricter immediately after than during the conflict itself because of the imperative of feeding the huge European population under Allied control.[b] Clothing was also rationed, on a point system. Two points were required for a pair of knickers,[c] five points for a man's shirt, five points for a pair of shoes, seven points for a dress, and 26 points for a man's suit. Clothing points could be traded for other utilitarian purposes, and Brenner eagerly traded with his landlady for a full English breakfast.

Perhaps on a whim, perhaps out of loneliness for former friends in South Africa compounded by the sudden loss of Harold Daitz, Brenner contacted May Covitz, now living with her son Jonathan in a small flat in Chelsea and juggling the dual roles of single parent and psychology student at London University. Jonathan Balkind recalls his mother asking him whether he remembered a man called Sydney who was once his baby sitter and friend. "She told me that he was also living in England, but in a different town."[3] Though deeply immersed in their work, the couple began socializing on a regular basis, most often in London. To May's surprise Sydney proposed marriage on one of these visits—"in the same offhanded way that he might suggest we go for a cup of coffee!"[4] Having recently experienced the emotional and logistical dislocation of a divorce, May was hesitant about remarrying. "One day I asked him to give me a mental picture of how he saw married life with me," she told a friend years later. "He replied, 'I see two people in a room with two desks and typewriters at opposite ends. Each of us has a window. We work at our desks and at some point I ask whether I should put the kettle on for tea. We chat for a while over tea and go back to our respective desks to carry on working. That is how I think about marriage with you.' 'Well sir,' I said, 'if that's your idea of marriage then I will marry you.' "[5]

On December 6, 1952, Sydney and May were married at a civil registry, each accompanied by two friends—and young Jonathan, of course. They treated themselves to a very brief London honeymoon. May recalled that they walked to their hotel with suitcases in tow through one of the worst fogs she could remember—"happy to be starting a new life together."[4] The

---

[b] Food rationing in Britain was finally discontinued in about 1954.

[c] "Knickers" originally referred to mens' garments, such as knickerbockers (also known as "plus two's" or "plus four's"). Nowadays the term is more frequently used in reference to underwear.

couple commuted back and forth between London and Oxford until May had largely completed her formal post-graduate studies in London. A few experiments and the arduous task of writing a thesis were all that remained to satisfy the requirements for her Ph.D. This consideration, encouraged by morning sickness that announced that she was pregnant (with a second son, Stefan, born in October 1953), prompted May to move to Oxford.

"Oxford wasn't especially interested in scientists in those days," Brenner commented. "Graduate students in the life sciences in particular were considered second-class. Additionally, being South African I gravitated to a lot of colonials and of course colonials were in the second division of the second class." In the chemistry department he met Jack Dunitz, a British X-ray crystallographer who had studied at California Institute of Technology (Caltech) before returning to Oxford for further fellowship studies.[d] The pair occasionally dropped in at Halifax House, conveniently located across the street from the Department of Physical Chemistry. Initially intended as a hostel, club, and social center for postgraduates and senior members of the university without college affiliation, Halifax House housed the Oxford University Graduate Club, a social club for men and women that provided facilities for social gatherings, reading, and writing.[6]

Through Dunitz, Brenner was introduced to Leslie Orgel, a young Englishman who had just completed postgraduate studies in chemistry at Oxford and was beginning his career as a theoretical inorganic chemist. The three became friends. At Brenner's suggestion, they initiated a discussion group in which Sydney expanded Orgel's and Dunitz's meager knowledge of basic genetics, and they, in turn, educated him in the basics of X-ray crystallography. "The first time that they mentioned Bessel functions[e] I thought they were talking about a Jewish wedding," Brenner wryly commented. Orgel and Dunitz were quickly caught up in Brenner's infectious enthusiasm about the mysteries surrounding the gene. "No hearing person around Sydney could remain unaffected by his enthusiasm, so I was inducted into subjects that were still unknown territory to almost everyone else studying chemistry," Orgel wrote.[7]

Under Hinshelwood's directorship, research in the Department of Physical Chemistry was intensely focused on chemical kinetics. In the years immediately

---

[d] Jack Dunitz had a highly successful career as a chemist, eventually taking up a professorship in Switzerland. He was elected a Fellow of the Royal Society (a rare event for a foreigner) and was made an Honorary Fellow of the Royal Society of Chemistry.

[e] Bessel functions, named for the mathematician Friedrich Bessel, refers to a mathematical term used in solving differential equations.

preceding Brenner's arrival, Hinshelwood had broadened his interest to embrace kinetic studies on the acquisition of bacterial resistance to a range of antibacterial agents and their adaptation to new sources of carbon and nitrogen as nutrients.[8] As we saw in Chapter 5, in his early correspondence with Brenner, Hinshelwood dropped the magic word "phage." To his satisfaction Brenner learned that the Oxford chemist's preoccupation with bacterial resistance embraced understanding why bacteria sometimes manifested resistance to infection by bacteriophages, viruses that infect bacteria. Not too long before leaving South Africa, Brenner had discovered a small book called *Viruses 1950*, which contained the published proceedings of a symposium on bacteriophages and viruses held at Caltech in March 1950.[f] "I knew about the work going on with phages by people like Luria and Delbrück and I was excited about the prospect of working with these bacterial viruses."

The Department of Chemistry was poorly equipped for biological studies, certainly for phage work, and Brenner had to scrounge bits of equipment and other essentials, even co-opting a domestic pressure cooker for sterilizing glassware. However, the Hinshelwood laboratory was adequately stocked with bacterial strains, and phage stocks were readily obtained from Sir Paul Fildes. This charming and eccentric member of the old school of English microbiology maintained a research group at the Sir William Dunn School of Pathology, conveniently situated across the street from the chemistry building.

Bacteriophage research is fundamental to the beginnings of molecular biology, and its early history merits a brief recounting. Discovered in 1915 by the Englishman Frederick Twort, and independently in 1917 by the French Canadian Félix D'Herelle, phages are viruses that infect and often kill bacteria. Early phage research was primarily concerned with defining the nature of these tiny infectious agents. The leading theories proposed that, like tobacco mosaic virus discovered some 20 years earlier, bacteriophages were either filterable viruses or some sort of self-perpetuating enzyme, the expression of which destroyed bacterial cells. The observation that bacteriophages can kill bacteria prompted attempts to use them as antibacterial agents, a notion presented in Sinclair Lewis's Pulitzer Prize-winning 1924 novel *Arrowsmith*. But research on "phage therapy" was abandoned in most of the Western world when penicillin and other chemical antibiotics were discovered.

The modern era of bacteriophage research dates from about 1938, when the expatriate German physicist Max Delbrück recognized the enormous

---

[f] The complete title of this volume is: *Viruses, 1950. Proceedings of a Conference on the Similarities and Dissimilarities between Viruses Attacking Animals, Plants, and Bacteria, Respectively.*

potential of exploiting these viruses to understand genes. A physicist by training and initially bent on a career in theoretical physics, Delbrück had a keen interest in biology that was awakened by the physicist Niels Bohr, with whom he worked in the early 1930s. But few biologists understood what Bohr was really getting at. As Delbrück bluntly put it, "[F]or them anything like quantum mechanics was utterly beyond their ken. At that time biologists didn't know any atomic physics. . . . [Furthermore], the biochemists at that time were super confidant that eventually everything would turn out to be biochemistry. . . . So this was Bohr's bold step, and it constituted for me the motivation to turn to biology."[9]

Reluctant to remain in his native Germany, which was moving ineluctably toward war, Delbrück secured a fellowship to join the *Drosophila* geneticist Thomas Hunt Morgan, who had recently moved his research program from New York to Caltech. But Delbrück found *Drosophila* genetics too complex for his taste. One day he learned that, while he had been away on a camping trip, Emory Ellis (a geneticist in Morgan's department at Caltech) had presented a seminar on bacteriophages. Disappointed at having missed the seminar, he sought out Ellis for a quick summary of his lecture and an opportunity for a quick primer on bacteriophage biology in general. Delbrück was impressed by how much information Ellis had collected using such modest tools as pipettes, Petri plates, and some agar. He was delighted to hear that such simple procedures allowed for the detection of individual virus particles with the naked eye. "I mean you could put them on a plate with a lawn of bacteria and the next morning every virus particle would have eaten a macroscopic 1 mm hole in the lawn. You could hold up the plate and count the plaques. This seemed to me just beyond my wildest dreams of doing simple experiments on something like atoms in biology."[9] Thus, Delbrück dropped *Drosophila* and teamed up with Ellis.

When Delbrück's fellowship expired in the late 1930s, he was faced with a serious dilemma. The war in Europe had begun, making it virtually impossible for him to return to Germany. But fortunately he was offered a teaching position in the Department of Physics at Vanderbilt University in Nashville, Tennessee, which allowed him to continue his fledgling studies with bacteriophage.

Salvador Luria, another physicist-turn-biologist, had independently discovered the wonders of bacteriophage biology in Italy. In 1938, Luria received a fellowship from the Italian government to work in the United States for a year and chose to spend it with Delbrück. Within a matter of days after Luria received his fellowship, Mussolini's Racial Manifesto went into effect and Luria was barred from activating the award because he was Jewish. Fleeing to France and to Portugal, Luria found his way to the United States and secured a faculty position at Columbia University. Luria and Delbrück finally met at the end of 1940, at the annual meeting of the American Physical Society in New York. The pair spent New Year's Day in Luria's laboratory and made plans to collaborate during the following summer at Cold Spring Harbor Laboratory (CSHL) on Long Island.[10]

The famous "phage group" (with Delbrück as its unofficial leader) evolved as an informal gathering of scientists from around the world who came to Cold Spring Harbor each summer, reveling in the opportunities for open discussion and a free exchange of ideas. The laboratory soon became an international phage Mecca, offering courses in phage biology and genetics. Delbrück was invited to return to Caltech in the mid-1940s as a member of the faculty, where he remained until his retirement. In 1969, Delbrück, Luria, and Al Hershey, won the Nobel Prize for Physiology or Medicine for their discoveries on the replication mechanism and genetic structure of viruses. (Hershey, with his student Martha Chase, demonstrated that the DNA of bacteriophages, not the phage protein, was the chemical agent responsible for generating phage progeny.)

Brenner fully appreciated the ease and power of using bacterial viruses as model organisms for exploring genes. As he devoured the relatively scant bacteriophage literature, familiarizing himself with the work of Delbrück, Luria, Gunther Stent, and others, he became excited at the opportunity to pursue phage studies in Hinshelwood's laboratory. However, he soon came to have fundamental disagreements with his mentor, who approached the phenomenon of phage resistance with a bias that galled Brenner's deeper understanding of bacterial genetics. Hinshelwood was not really interested in genetics and frankly did not believe in mutations, claiming that mutagenesis was not a rigorously established phenomenon. Further, he maintained that phage resistance in bacteria was an adaptive rather than a genetic phenomenon. Luria, who conceived the famous Luria–Delbrück experiment, the "fluctuation test"[g] that formally demonstrated the phenomenon of

---

[g] The Luria–Delbrück fluctuation test is a classic statistical analysis of bacterial mutagenesis that was designed and executed by its famous authors in 1943. The work lent enormous credence to the notion that mutations in bacterial (and presumably other) cells can arise spontaneously, and do so stochastically. The test is still sometimes used to demonstrate the phenomenon of spontaneous mutagenesis.

mutagenesis in bacteria, addressed these assertions:

> Traditional wisdom among bacteriologists in those days had it that
> bacteria had no chromosomes and no genes. This idea was bolstered by
> the authoritative opinion of an eminent British physical chemist, Sir Cyril
> Hinshelwood, who through mathematical models explained all hereditary
> changes in bacteria as due solely to altered chemical equilibria. I have
> often noticed in later years that biologists are readily intimidated by a bit
> of mathematics laid before them by chemists or physicists.[11]

Brenner thus felt obliged to repeat the fluctuation test to prove defini-
tively the phenomenon of mutagenesis to Hinshelwood. "I'm probably the
only scientist in the world who was prompted to repeat this fluctuation test
for his Ph.D. thesis," he lamented. Brenner believed, as did much of the phage
world, that whereas phage resistance was the result of a mutation in the
bacterial genome that interfered with some step in the process of phage infection, lysogeny (a state in which the phage genome is quiescently integrated
into the host genome) was, indeed, an adaptive phenomenon. Brenner tried
hard to persuade Hinshelwood to let him work on lysogeny.[h] The pair had
many heated (though generally friendly) discussions about this fundamental
distinction. One day in a pique of intense frustration, Brenner blurted to his
mentor, "Professor Hinshelwood, in science, as in life, it's important to distinguish between chastity and impotence. The outcome is the same but the reasons are fundamentally different. One state arises because one can't and the
other arises because one won't. Phage resistance derives from a mutation
that *can't* permit the phage to infect bacteria. That's impotence! Lysogeny
arises because the phage *chooses* to remain in the vegetative cycle. That's
chastity!"

No one in the Physical Chemical Laboratory at Oxford argued with
Hinshelwood and some were taken aback by Brenner's outburst. But in the
main the two enjoyed a comfortable relationship. Hinshelwood was actually
an attentive listener, always willing to consider well-presented arguments.
In the end he conceded that some changes in microorganisms might, indeed,
be due to mutations. Almost 30 years later, when reviewing Salvador Luria's
autobiography, Brenner pointedly noted that Luria and Delbrück invented the
fluctuation test to refute the views of Sir Cyril Hinshelwood, who did not
believe in genes or mutation but only in chemical equilibria. "I happen to

---

[h] In the lysogenic (vegetative) state the phage genome is dormant until some event triggers its
dissociation from the host genome and its autonomous replication to generate more phages
(the lytic state).

know it did not convince him at all, because I spent several months at Oxford in 1952 and 1953 doing enough fluctuation experiments to prove to Sir Cyril that the results could not be explained by dirty test tubes."[12]

In the course of his studies, Brenner isolated numerous phage mutants, many of which had nothing to do with bacterial resistance. His curiosity led him to study some of them more deeply. For example, when experiments yielded mutants defective in tryptophan biosynthesis, he set about synthesizing intermediates in this process to comprehend this phenotype more precisely. No one in the chemistry department knew much about tryptophan biosynthesis, so Brenner sought help in the nearby Dyson-Perrins Laboratory that housed the Department of Organic Chemistry. One morning he was seated at his borrowed laboratory bench when Sir Robert Robinson[i] swept in and asked who he was and what he was doing. "I told him what I was up to and he appeared genuinely interested. Of course I didn't tell him that I was an interloper in the laboratory. But apparently about a week later he stopped by the lab again and asked the whereabouts of that interesting young man."

Shortly before Brenner left South Africa, Wits principal Raikes contacted other former Oxford acquaintances on his behalf. These included Richard Dawkins, professor of Greek studies.[j] "My dear Dawkins," Raikes had penned, "You must look after Sydney Brenner who is to work with Hinshelwood as a 1851 Exhibition Scholar."[13] Brenner was captivated by Dawkins, who often invited him to tea in his rooms in Exeter College. He was especially intrigued to learn that Dawkins knew A. J. A. Symons, author of the English classic *The Quest for Corvo,*[k] a marvelous account of a particular social culture in England about which Brenner had read in South Africa, presumably in anticipation of his move to Oxford.

In between work in the various laboratories and informal meetings with Leslie Orgel and Jack Dunitz, Brenner squeezed in as many seminars and lectures as he could, especially from visiting notables. He was especially excited to hear Fred Sanger, now no longer a graduate student, who came over from

---

[i] Robert Robinson (1886–1975) was a renowned Oxford organic chemist and Nobel Laureate in Chemistry in 1947 for his work on anthrocyanins (plant dyestuffs) and alkaloids. In 1953 the American Chemical Society awarded him its highest prize, the Priestley Medal.

[j] Richard MacGillivray Dawkins was the first Bywater and Sotheby Professor of Byzantine and Modern Greek Language and Literature at the University of Oxford. See http://en.wikipedia.org/wiki/Richard_MacGillivray_Dawkins.

[k] Corvo was a pseudonym for Frederick Rolfe, one of English literature's great eccentrics and the author of *Hadrian the Seventh.*

Cambridge to present his pioneering work on protein sequencing. Protein structure had been a total mess until then. Linus Pauling and Robert Corey had worked out the nature of the peptide bond, but no one had persuasive ideas about protein structure. In fact many simply thought of these macromolecules as formless blobs. But Sanger showed that different proteins had different amino acid sequences, suggesting different three-dimensional structures. "He used little wooden blocks to demonstrate this to his audience," Brenner recalled. "It was a phenomenal presentation. At the end of his talk Sir Robert Robinson stood up and announced: 'Doctor Sanger has now made proteins a part of chemistry.' " This busy schedule presumably cut into his time with May, Jonathan, and their baby son Stefan. But May was no stranger to her husband's passion for science. Besides, she was a professional too.

Through his acquaintance with Dunitz, Brenner met Jerry Donohue, a member of the X-ray crystallography group assembled by Lawrence Bragg at the nearby Cavendish Laboratories at Cambridge. Donohue's visits to Oxford afforded further opportunities for talking about science, frequently in the physical chemistry library. In January of 1953, a few months before Watson and Crick published their historic *Nature* paper announcing their model of the DNA structure, Brenner and Donohue were chatting about DNA. At one point in the conversation, Brenner sketched the structure of one of the pyrimidines on a piece of paper, prompting Donohue to ask why he drew the base in the rare enol form. Brenner casually responded that he had always drawn the nitrogenous bases in DNA that way and the discussion proceeded. Donohue had made the same observation to Watson, a point that was crucial to Watson's success in correctly pairing A with T and G with C forms to model base pairing in DNA.[1]

At that meeting Donohue informed Brenner that Watson and Crick had assembled a viable model of the DNA structure. Indeed, soon thereafter he arranged for a group including Dunitz, Brenner, Orgel, and Beryl Oughton (a student in Dorothy Hodgkin's laboratory at Oxford) to view the model

---

[1] While struggling to come up with a structurally plausible scheme for base pairing in DNA, Jim Watson constructed large cardboard shapes of the bases A, T, C, and G and tried to pair them in a manner that would yield consistent dimensions, a necessary feature if the base pairs were to be incorporated in a coherent DNA structure. Jerry Donohue happened to be in Watson's office one day and pointed out to him that even though all the textbooks showed the chemical configuration of the bases with which Watson was toying, he was, in fact, using the rare tautomeric forms. Once Watson switched to cardboard shapes that reflected the more common tautomers, he found that A:T and G:C base pairs had the same dimensions. This was a pivotal turning point in the elaboration of a correct model of the DNA structure.

first-hand. The group drove to Cambridge on a cold spring morning in April 1953. "We entered a brick-walled room in the Austin wing of the Cavendish Laboratories and there stood the model; connected with tall laboratory retort stands and clamps, the bases represented with shiny metal plates that had been specially machined to scale."

Brenner was spellbound. Viewing the model was nothing short of an epiphany for him. Seeing the organized helical structure with the bases neatly paired in strict purine-pyrimidine fashion, he was overwhelmed by the intuitive understanding that a structure so well defined ("beautiful" was the word that Watson repeatedly used later) was going to be enormously informative about genes. "I suddenly realized that the simplistic ideas that I had been bumbling and fumbling with about coding and protein synthesis could now be tackled in a meaningful way. Standing there and looking at the DNA structure I felt as if a curtain had been lifted in my mind, in the sense that I knew that this was a fundamental breakthrough in biology and that we could now find out how genes worked. On that day I knew exactly what I wanted to do. I wanted to work on genes and their function."

Twenty-one years later, when writing a piece for *Nature* in celebration of the "coming of age" of DNA, Brenner still waxed lyrically.

> The double helix fundamentally changed the image of biology. To most young people of my generation the biology taught in universities was a most unattractive subject. It seemed to consist in learning long dusty lists of Latin names punctuated by cutting up frogs or carrots in long dusty laboratories. DNA changed all that and turned biology into an exciting, intellectually attractive, subject.[14]

Brenner spent much of that day talking with Watson and Crick. In particular, he and Watson took a long walk around the university campus, chatting about their mutual interest in bacteriophage biology (Watson had worked with phages when he was a graduate student with Luria and during his post-doctoral fellowship) and about the implications of the DNA structure. Brenner's immediate impressions of Watson were of a bright but rather eccentric person who walked rapidly with long strides and who did not take as much notice of him as he would have liked. Commenting on this walk again years later he wrote:

> [H]e strode out at a tremendous pace and I almost had to run to keep up with him. He left the impression with me of a large irritated bird stamping about, pecking here, there and everywhere. Walking around Cambridge we talked about some of the biological implications of the DNA structure.[15]

Watson remembers his first encounter with Brenner. In an introduction to a 1983 conference to celebrate the 30th anniversary of the discovery of the DNA structure, he wrote:

> On the first occasion I saw Sydney, we talked for about six hours non-stop. . . . He took me aside . . . and said: "Jim, you don't realize how important the work you've done is." I think I did, but I was also scared it might not be right. [Sydney] was clearly very bright. He had spent much of his time in South Africa reading the scientific literature, so he was extraordinarily well informed. And he also read a lot while he was at Oxford. So his knowledge base was quite phenomenal. I remember too his great sense of humor. He was fun to be with.[16]

On that occasion Brenner enjoyed less interaction with Francis Crick, who in Watson's words "was happily giving his oration on the DNA structure."[16] "I had heard a lot about Francis and when I met him that day in Cambridge his enthusiasm and energy impressed me immediately," Brenner related. "He couldn't stop talking about the model and its implications and it was enormously interesting to listen to him prattle away in his loud voice."

The implications of Watson and Crick's model did not immediately take the scientific world by storm. As noted by science historian Robert Olby, the double helix had a rather "quiet debut" in the spring of 1953.[17] Many explanations have been advanced to explain this lapse of comprehension. Most notably perhaps, molecular biology, operationally defined then as the synthesis of biochemistry and genetics, was not yet a formalized discipline. The great biochemists of the day, such as Fritz Lipmann and Theodor Lynen in the United States, and Hans Krebs in the United Kingdom, were still industriously (and profitably) mining the powerful gains of enzymology and unraveling the intricacies of intermediary metabolism. Even leaders in the world of protein synthesis, such as T. S. "Tommy" Work in England, were not convinced about a fundamental role for nucleic acids in this process. To be sure, in a review article on the *Biosynthesis of Proteins* published in *Nature* just a few months after Watson's and Crick's report, Work and Peter Campbell wrote:

> The conception of the gene is essentially an abstract idea and it may be a mistake to try to clothe this idea in a coat of nucleic acid or protein. . . . If we must have a gene it should have a negative rather than a positive function as far as protein synthesis is concerned. A more prevailing theory held that proteins were synthesized by the formation of a succession of peptides, ultimately yielding the peptide molecule.[18]

This so-called peptide model of protein synthesis was supported by several prominent biochemists of the day. Brenner expounded:

> The idea that you could glean anything from the DNA structure that would tell you something important about biology didn't catch on immediately. . . . Furthermore, the biochemists were generally skeptical about genetics. When I was doing phage experiments at Oxford someone once asked me what statistical methods I used to validate my results, something that was of course standard in biochemistry. I facetiously told him that I didn't have to do statistics. If I was testing for recombination and used mutants that were non-revertible, either I got a recombinant or I didn't. Later on I glibly told this person that we sometimes plotted our results on seven-cycle log paper and held up the graph at one end of the room. If one could see a difference from the other end of the room it was significant!
>
> But the fact is that in 1953 the necessary fusion of biochemistry and genetics had not yet taken hold. . . . Indeed, people like Erwin Chargaff[m] considered molecular biologists to be nothing more than a bunch of biochemists who practiced without a license.

A similar attitude emerged a few years later when Crick formulated his adaptor hypothesis for the assembly of amino acids in a growing polypeptide chain and suggested that there would be 20 different enzymes to join the transfer RNA to its cognate amino acid. "The biochemists said that this was utterly impossible, because if there were twenty such enzymes they would have already found them!" Brenner stated.

To a large extent this attitude was compounded by the reality that genetics was in something of a mess in the mid-20th century. Morgan was an embryologist, not a geneticist. He initially turned to *Drosophila* genetics in the hope of understanding development, not the action of genes. Ironically, he and his disciples realized more significant gains in genetics than in development. Even when classical genetics was more firmly established, it was initially purely descriptive. Indeed, when Morgan and his student Alfred Sturtevant extended their studies on the linkage of genes to mapping their relative distance, the analogy emerged of the hereditary material as a string of beads. However, deciphering what these beads consisted of and how they functioned remained a baffling mystery. As Horace Freeland Judson noted, "[S]mall wonder that in his 1933 Nobel lecture Morgan put aside consideration of the physical nature of the gene as unnecessary or premature. To get any

---

[m] Erwin Chargaff, a renowned nucleic acid biochemist at Columbia University, is perhaps most famous for having discovered that the ratio of purines (A and G) and pyrimidines (C and T) in DNA is identical.

deeper into the gene than they did, with the system they had, was then not technically possible."[19]

Given this history it is not surprising that between 1953–1960 scientific publications on DNA were mainly concerned with methods of its extraction, its physical properties, and whether the content and composition of DNA was the same for all cells of a given organism. Only cursory mention was made in the literature of the implications of its structure for understanding the replication and coding properties of DNA. Correspondingly, the only documented public acknowledgement of Watson and Crick's discovery was by way of a small piece in the *News Chronicle*, a British daily newspaper. With Queen Elizabeth II's grand coronation and Edmund Hillary's conquest of Mount Everest, British newspapers had much more exciting stories to cover in the spring of 1953.

The two DNA chains were not simply parallel coiled; they were intertwined. So many who examined the model were confounded by the perceived complexity of replicating such a molecule. "This was the first time that I adopted what I believe to be an important view about models and hypotheses in general," Brenner said. "I later called it the 'Don't Worry' hypothesis, meaning that if an idea or a model is compelling, don't worry too much about details that cannot be immediately explained. It can be useful to lessen the constraints of a model or a theory, provided of course one doesn't ignore solid facts." The folding of polypeptide chains into proteins provides a fine example of this hypothesis. Most biologists thought it inconceivable that proteins could correctly fold on their own and tacitly assumed that specific genes were devoted to folding polypeptide chains. Brenner's view was simple and straightforward: "There's probably some kind of enzyme to unwind the DNA chains," he predicted. He was correct, of course, even though no one then knew of DNA helicases.

Brenner's view of the way DNA went about its business was distinctly one-dimensional. In his mind this simplification made phenomena such as DNA replication and gene expression more comprehensible. "I had no difficulty in thinking about DNA as a one-dimensional functional entity because I had read John von Neumann about computers, which he discussed in terms of uni-dimensional sequences. So instead of thinking about the problem of gene coding one could pose the simple question: 'If I understood a gene unidimensionally, that is, in terms of its nucleotide structure, and I could also sequence the protein it encodes, might I show that they were strictly collinear?'" This crucial question was to occupy much of Brenner's intellectual focus for years to come.

In a little more than two years, in 1954, Brenner had completed a doctorate thesis entitled "The Physical Chemistry of Cell Processes: A Study of Bacteriophage Resistance in *Escherichia coli*, Strain B12." Excerpts from the abstract of his thesis indicate his conclusion that phage resistance is determined by mutations—not adaptation:

> There has been much controversial discussion on the origin of bacterial variants. Such new strains appear readily when populations of bacteria are exposed to new environmental conditions, and two theories have been proposed to explain this phenomenon. The first, the theory of mutation selection, postulates that the variants arise by spontaneous genetic changes which occur independently of the new environment. . . . The second, the hypothesis of adaptation, states that the variants arise by a direct interaction of the cells with the new environment, which induces modifications in the metabolic properties of the cells. . . .
>
> . . . The origin of the phage resistant cells in *E. coli* B has been investigated using fluctuation tests and replica plating. . . . It has been concluded that physiological and environmental differences do not significantly affect the determination of T7 resistance and that the fluctuations found are consistent with the mutational origin of the resistant cells.[20]

Max Delbrück's remark about Watson's thesis when he was a graduate student with Luria—that it was sound, but sufficiently boring that he (Watson) would avoid the trap of pursuing his thesis research further, freeing him to seek new horizons—was perhaps also true of Brenner's thesis. Importantly, Brenner fully comprehended the virtues of phage biology and had met two individuals who would soon be credited with the most fundamental contribution to biology since Darwin and Mendel. He was now poised and eager to launch his own career in molecular biology.

Brenner was then faced with an important logistical decision: where and how to pursue his academic future. With the impending termination of his 1851 Exhibition Award, he began seriously to think about returning to South Africa, as required by the terms of the award, and to accept the closure of this chapter in his life. A return to the country of his birth was vexing for both Sydney and May. The political climate in South Africa showed no indication of improvement. On the contrary, the conservative and abjectly racist Nationalist Party that surged to power in the late 1940s was rapidly consolidating its position on "separate development" of the country's minority white and majority non-white populations. The new government certainly had no plans to heighten South Africa's limited research profile in the world, and the prospects of becoming a big fish in a very small scientific pond held little appeal for Brenner.

Brenner and May talked often about the wisdom or folly of returning to South Africa. Prompted by his pledge to honor the terms of the 1851 Fellowship and reinforced by the knowledge that Joe Gillman, now professor of physiology at Wits, was eagerly anticipating Brenner taking a junior faculty position in his department, the couple reluctantly decided to pursue this course. At the eleventh hour, however, Brenner reconsidered this decision after notification of a position in the United Kingdom. The offer arose through the intervention of one of Jim Watson's colleagues and friends, Murdoch Mitchison,[n] to whom Watson had told of the young South African. In the late summer of 1953, Brenner wrote to Watson:

> I . . . had a letter from Murdoch Mitchison who mentioned that you had spoken to him about me. I went across to Cambridge to see him and he has suggested that I come and work in Edinburgh next year. His offer is quite attractive and I am seriously considering it, but, at the moment, I'm terribly confused about what I want to do (or ought to do) when I've eventually extricated myself from Oxford.[21]

The scope of Brenner's self-admitted confusion about his professional future is revealed in other opportunities he was then quietly contemplating.

> I saw Jack [Dunitz] who is leaving for the States next week via Canada and he reiterated his suggestion that I should go to Cal Tech. It may well be possible that I shall have the chance of being able to go to the States for a year by myself (I doubt there'll be any difficulties re visas) and leave my family in England. Do you think it would be possible for me to come and work in Pasadena [where Watson was doing post-doctoral work with Delbrück] for a year starting in July 1954? You did mention at one time that I should go to Cal Tech but until recently I have never seriously considered it. If you have any suggestions to make on this matter I would be glad to hear them, because I want to make a final decision by the end of the year.[21]

These practical dilemmas aside, the two and a half years that Brenner spent in England represent a fundamental turning point in his life. In his own words:

> When I went to Oxford in October 1952 to work on bacteriophage with Hinshelwood it was with the intention of seeing whether physical chemistry

---

[n] Murdoch and Avrion Mitchison are sons of the writer Naomi Mitchison, to whom Watson dedicated his famous nonfiction novel *The Double Helix*. Watson was best man at Avrion Mitchison's wedding.

could provide help in solving biological problems. I should have gone to study molecular biology, but the subject did not yet exist. From my past experience in cytology and cytogenetics I knew that DNA was the material basis of heredity and that RNA was important for protein synthesis. ... I had read von Neumann's article on the theory of self-reproducing machines. Beyond this I had many nebulous ideas on how nucleic acids might exert their function and on how we might test them, including one ridiculous proposal that the structure of nucleic acids could be solved by dichroism measurements of DNA complexed with dyes.

On a chilly morning in April 1953 ... I went to Cambridge and saw the model and met Francis and Jim. It was the most exciting day of my life. The double helix was a revelatory experience; for me everything fell into place and my future scientific life was decided there and then.[22]

7

Confronting the Genetic Code

There was a way to eliminate all overlapping triplet codes

IN THE AUTUMN OF 1953, MILISLAV DEMEREC, director of the Cold Spring Harbor Laboratory on Long Island, New York, visited Hinshelwood. He was introduced to Brenner, who enthusiastically communicated his work on bacteriophage resistance in bacteria. Demerec was impressed and immediately suggested that Brenner consider visiting Cold Spring Harbor before returning to South Africa. Under Demerec's guidance Brenner applied for financial support from the British Dominions and Colonies Fund of the Carnegie Corporation of New York. Once again the recipient of Andrew Carnegie's largesse in support of English-speaking countries in the world, Brenner was awarded a grant of $2100 to cover the cost of round-trip travel from England to the United States and a sojourn in the United States for four months. In his application Brenner announced his intention to spend ten weeks at the Cold Spring Harbor Laboratory, during which he planned to participate in the annual phage course. He also proposed visiting other research facilities in the United States, including laboratories in New York, Philadelphia, New Haven, Baltimore, Bethesda, Chicago, Caltech, Berkeley, and Stanford. He excitedly communicated his good news to Watson in hopes of visiting him either at Cold Spring Harbor or Caltech.

Some in South Africa, most notably Humphrey Raikes, were unsettled at the prospect of Brenner's planned six-month divergence to the United States. His erstwhile supporter wrote:

> While I have supported your application to Carnegie very strongly, I must frankly confess that I am a bit alarmed at this sudden change in your arrangements. As you know, I have the greatest possible regard for your mind, but I do feel that you still jump about more than you ought and do not keep to a steady path.[1]

Brenner lost no time in responding to Raikes, steadfastly defending his decision.

> I received the impression from your letter that you feel that this projected trip to America is a sudden change in my arrangements. I naturally intend completing my D. Phil. at Oxford before leaving and hope to have done so by the end of the summer term. ... There is very little work done in this field in England and I am the first person to do such work in Sir Cyril's laboratory. Most of the work is being carried out in the United States, and when Dr. Demerec was here he suggested that I should go to America for further study. ... The proposed four months visit in America is intended to consolidate my knowledge in this field in which I have worked for two years.[2]

The question as to whether May and his stepson Jonathan would accompany Brenner to the United States was never seriously entertained. Her decidedly leftist political views were of obvious concern, especially in light of the intense anti-Communist sentiments in the United States then. Senator Joe McCarthy, the vitriolic "red-hater," was at the peak of his communist witch hunts. Even the American chemist Linus Pauling was denied a visa to visit the Cavendish Laboratories in the United Kingdom because of his political activism. Furthermore, having lived in the United States with her first husband during World War II, May Brenner was not enchanted with the country and not especially eager to return. "So my family returned to South Africa and I went off to Cold Spring Harbor for the summer of 1954."

Prior to his visit to Cold Spring Harbor, Brenner (like Watson and Crick) had been struck by a theoretical paper published in *Nature* in early 1954 in which George Gamow, a renowned Russian theoretical physicist and cosmologist (who made, among other notable achievements, seminal contributions to the "Big Bang" theory on the origin of the universe), advanced a madcap theory of how the genetic code might work. "Gamow's theory was hopelessly wrong. But it got a lot of people, including me, thinking about a DNA code for specifying protein synthesis."

Aptly described by Watson as "the truly bizarre explorer ... theoretical physicist extraordinaire, and a six-foot, six-inch giant,"[3] Georgy Antonovich Gamow, referred to in the English-speaking world as George or Geo (pronounced "Joe") "because that's what Gamow insisted was the correct pronunciation of Geo, ... defied conventional description with his aptitude for tricks that masked a mind that always thought big."[4] Gamow received his Ph.D. in physics from the University of Leningrad in the late 1920s and completed a postdoctoral stint at the Institute of Theoretical Physics under the legendary Niels Bohr.

Despondent at the increasingly oppressive political climate in the Soviet Union, Gamow and his wife Lyubov Vokhminzeva (nicknamed Rho—and also a physicist) decided to flee the country in 1932. Indicative of the essential gestalt of the man, Gamow and his wife attempted to cross the Black Sea to Turkey—in a collapsible rubber kayak no less! The Black Sea is infamous for its violent storms, and Gamow and his wife were blown back to shore after a long and miserable night on the ocean.[5]

About a year later Gamow was invited to attend the 7th Solvay Conference on Physics in Brussels. Though suspect of his political allegiance to the Soviet Union, the Russian authorities were proud of Gamow's professional fame and the kudos it brought to the country. Thus, they considered this invitation favorably. However, Gamow recognized that permission to travel to Brussels would provide a golden opportunity to defect. The problem was how to persuade the authorities to allow his wife Rho to accompany him. He conceived an ingenious ploy. He began petulantly announcing that if his wife were not granted a passport he would stay home. In due course a meeting was arranged at the Kremlin with the Communist Party Chairman Vyacheslav Molotov. In his autobiography, *My World Line*, published in 1970, Gamow related the outcome of his pivotal meeting with Molotov.

> "You see," I said, "to make my request persuasive I should tell you that my wife, being a physicist, acts as my scientific secretary, taking care of my papers, notes, and so on. So I cannot attend a large congress like that without her help. But this is not true. The point is that I want to take her to Paris to see the Louvre, the Folies Bergere, and so forth, and to do some shopping."[6]

The pragmatic Molotov readily identified with this argument and acquiesced to Gamow's request. In time (though not without several delays prompted by Molotov's subsequent vacillations), Gamow and his wife received passports and left Russia forever. Gamow became a naturalized American citizen in 1940.

Though lacking formal training in the discipline, Gamow was not totally naïve about biology. In 1953 he published an amusing popular book on biology entitled *Mr. Tomkins Learns the Facts of Life.*[a] Additionally, his stay in Niels Bohr's Copenhagen laboratory in the late 1930s coincided with

---

[a] Gamow wrote a series of books designed to popularize physics. One of these, called *Mr. Tomkins in Wonderland*, published in 1936, featured a fictitious character (Mr. Tomkins) who lives in a wonderland where buses and motor cars travel near the speed of light. Mr. Tomkins reappeared in 1947 in *One, Two, Three—Infinity* and in 1953 in *Mr. Tomkins Learns the Facts of Life*.

Delbrück's. The two physicists worked side by side, and some of their discussions presumably concerned the biological topics that then aroused Delbrück's interest.

One day in early June 1953, Gamow retrieved the mail from his residential mailbox in Bethesda, Maryland. He casually began perusing his May 30th issue of *Nature*. Watson and Crick's second 1953 letter to *Nature*, entitled *Genetical Implications of the Structure of Deoxyribonucleic Acid*, immediately caught his eye. The paper published almost exactly a month earlier is historically more famous. But it was the May 30th article that first announced some of the important biological implications of the DNA structure. In particular, the following decisive statement by Watson and Crick grabbed Gamow's attention.

> The phosphate-sugar backbone of our model is completely regular, but any sequence of the pairs of bases can fit into the structure. It follows that in a long molecule many different permutations are possible, and it therefore seems likely that the precise sequence of the bases is the code which carries the genetical information.[7]

Apparently, Gamow was so intrigued by the mathematical implications of a code embodied in the millions of iterations of the four DNA bases A, C, T, and G that he read the entire article while standing at his mailbox.[8] Gamow lost no time in introducing himself to Watson and Crick with a zany handwritten letter dated July 8, 1953:

> I am a physicist, not a biologist and my interest in biology can be justified, if anything, by my recently published book 'Mr. Tomkins Learns the Facts of Life' (Cambr. Univ. Press). But I am very much excited by your article in May 30th Nature, and think that this brings biology over into the group of 'exact' sciences.[9]

The theoretical physicist proposed an elaborate coding scheme (which he subsequently published) that became known as "the diamond code." This, as well as several other coding schemes that Gamow proposed (all rather ingenious), turned out to be utterly wrong. Important, however, is the fact that he stimulated others (including Brenner) to think seriously about the enigma of genetic coding, and in so doing undeniably accelerated its solution.

A central feature of Gamow's diamond code is that it addressed the problem that there were 20 known amino acids and only four bases. A code comprising two bases as a coding unit came up short because there are only 16 possible doublets of the four bases. Using three bases as a coding unit readily accommodated all the amino acids, even though there was the disquieting realization that there are 64 possible triplet combinations of the four bases—more than three times the number needed to encode 20 amino acids. (Degeneracy in the genetic code was a revelation that still lay ahead.) Nonetheless, Gamow (correctly) invoked a triplet code for the purposes of his model and suggested that each triplet overlapped by one base, thereby yielding exactly 20 coding units.

Notwithstanding the intuitive appeal of a model that accounted for the 20 known amino acids, as soon as Watson and Crick examined Gamow's coding scheme it became obvious to them that his world-class skill as a theoretical physicist did not translate to the world of biology. Gamow's model was frankly impossible. Aside from the fact that he completely ignored the role of RNA in protein synthesis, he offered no chemical explanation as to how particular amino acids recognized and slotted into the postulated diamond-shaped cavities in DNA.

Disposing of some of the specifics of Gamow's wacky code was one thing. But disposing of the general concept of an *overlapping* genetic code was an entirely different matter. Having read Gamow's February 1954 *Nature* paper (entitled Possible Relation Between Deoxyribonucleic Acid and Protein Structures), Brenner began to think about the feasibility of overlapping codes and how that notion might be proved or disproved before he went to Cold Spring Harbor. He came up with an ingenious theoretical experiment that required nothing more than pencil and paper—and access to sufficient information about known dipeptide sequences in proteins. Brenner offered the following:

> Let's imagine the hypothetical sequence ACGTA in DNA. If the triplet ACG specifies a particular amino acid, then in an overlapping code with a single nucleotide overlap the second triplet obviously has to be GTA. In a code with just four letters (ACGT) the number of possible overlapping adjacent triplet codons is 256 (4^3). But if the code specifies twenty amino acids, the possible number of *dipeptides* [author's italics] is 400 (20^2).

Brenner was arguing that, if there were, indeed, an overlapping triplet code, some dipeptides would be impossible, because overlapping codes would introduce restrictions in amino acid sequences. Thus, if one were to examine enough peptide sequences and demonstrate that all 400 possible dipeptides were, in fact, represented, that would eliminate an overlapping code.

"Sydney has this extraordinary ability to cut through a lot of what might be called 'stuff,' . . . the stuff that all scientists carry around in their heads and that frequently misdirects otherwise smart thinkers," Stanford biochemist

Paul Berg stated.[10] "I think that the word 'genius' is overused with reference to scientists," Berg continued. "But it's not inappropriate to use that word in relation to Sydney. The breadth of what his mind is capable of wrapping around is quite extraordinary. Additionally, he is blessed with the vision and courage to go out there and test ideas in truly original and creative ways."[10] Terry Sjenowski, a leading scientist who worked closely with Brenner for over a decade, has stated:

> Sydney is without question a genius; and not necessarily in the narrow sense that the word is sometimes appropriately used. He has a generalized genius that extends to the entire body of biology. He has a unique way of organizing a body of knowledge and he thinks about things in ways that are different from anyone else I have encountered—including Francis Crick. Both Francis and Jim Watson secured their places in history with their contributions to the DNA structure, probably the greatest discovery in biology in the 20th century. And both are very astute thinkers. But Sydney's influence on what happened subsequently is really quite phenomenal.[11]

Brenner began documenting peptide sequences whenever he could lay his hands on them. In the mid-1950s the number of known dipeptide sequences was considerably less than 400. Nonetheless, he determined that that number was sufficient to offer a statistical proof of the impossibility of an overlapping code. Simply stated, Brenner was able to demonstrate that, whereas the maximum number of possible triplet codons is 64, it would require more than that number to account for the data he had accumulated. A letter to Jim Watson written in late July, 1954 is revealing:

> Dear Jim
>
> Will Gamow still be at WH [Woods Hole] when I come up [from Cold Spring Harbor]? I have been looking into some of the implications of his model and there appear to be some difficulties. One of these is that his model predicts that each amino acid in a chain can only be followed by one of four amino acids, ... The data I've collected on amino acid sequences show that this is not the case; thus in insulin itself, glutamine is followed by 6 different amino acids.[12]

Remarkably, Francis Crick had come to the same conclusion before leaving Cambridge in August 1953 to take up a year at the Brooklyn Polytechnic in New York. He had been eagerly following Fred Sanger's progress in deciphering the amino acid sequence of insulin. When he returned to Cambridge in late 1954, Crick began preparing a manuscript for an entity called the RNA Tie Club (the origins of which are described presently) that offered the same conclusions about overlapping triplet codes that Brenner had. Crick and Brenner presumably discussed these notions when they renewed their acquaintance at Cold Spring Harbor, and perhaps more substantively at the Marine Laboratories at Woods Hole, Massachusetts, in the summer of 1954. Notably, however, Brenner proved the impossibility of *all* overlapping triplet codes, a feat that he, too, announced in an RNA Tie Club manuscript—in September 1956.

Initiated in the spring of 1954, the RNA Tie Club (and the related RNA tiepin) was one of George Gamow's more enterprising schemes. Inspired by his newfound interest in DNA and genetic codes, Gamow founded an informal "science club" that distributed new experimental observations, theories, and hypotheses among a select membership of budding molecular biologists. The emphasis on RNA rather than DNA presumably reflects the widespread acceptance at the time (even by Gamow) that RNA was somehow centrally involved in protein synthesis. In fact Crick was already formulating his famous "adapter hypothesis" that postulated the existence of multiple transfer RNAs in cells (he called them adapters). In fact the RNA Tie Club piece, called "On Degenerate Templates and the Adapter Hypothesis" and distributed to the club in early 1955, is more famous for is announcement of this hypothesis than for Crick's conclusions about nonoverlapping triplet codes.

The RNA Tie Club emblem was a black tie garishly embroidered with a green and yellow cartoon of an RNA molecule. Gamow ambitiously wanted to have the tie (prominently featured on the dust cover of the English edition of Watson's memoir, *Genes, Girls and Gamow*) produced by an up-scale haberdasher in Oxford. But he finally settled for a Los Angeles production. He proposed 16 members, a number soon expanded to 20—one for each amino acid. Each member was also promised a club tiepin that carried the three-letter abbreviation of his assigned amino acid. Four honorary members (including Watson and Crick) were also appointed by Gamow—one for each base. The membership was largely drawn from scientists that Gamow knew or knew about, who were interested in the coding problem, including several of his physicist friends, notably the theoretical physicist Edward Teller. An RNA Tie Club letterhead was generated that listed the club officers: Geo Gamow, synthesizer; Jim Watson, optimist; Francis Crick, pessimist; Martynas Ycas, archivist, and Alex Rich, Lord Privy Seal. The letterhead also bore the inspirational motto: "Do or die, or don't try," a slogan apparently suggested by Max Delbrück.

The primary goals of the RNA Tie Club (considered by some to be frankly lunatic—even for George Gamow) were to encourage discussion and to circulate unpublished manuscripts that were "more speculative, discursive, and untested than their authors would risk in formal publication."[13] As it turned out, only a handful of scientific papers were written and distributed. Nonetheless, for a brief period, the club served the purpose of disseminating unpublished information on the emerging discipline of molecular biology.

"During the period 1953–1954, there were very few people who even knew what we were talking about," Brenner said of the RNA Tie Club. "You can see that when you examine who Gamow had in his club. Francis, Leslie Orgel and I were the last three members. Francis was Tryptophan, Leslie was Tyrosine and I was Valine, the last member. Most of the rest were Gamow's friends. So Nick Metropolis was a member because he worked on computers at Los Alamos. Of course Gamow named himself after the alphabetically first amino acid, Alanine."

After Brenner submitted his piece on the impossibility of overlapping triplet codes to the RNA Tie Club, Gamow recognized that the contribution merited exposure well beyond that in-group. He, therefore, communicated it to the Proceedings of the National Academy of Sciences for formal publication with the expanded title: *On the Impossibility of All Overlapping Triplet Codes in Information Transfer from Nucleic Acids to Proteins.* He wrote to Brenner:

> This is just to let you know that I have mailed your article for publication in P.N.Ac.Sc. Of course, everybody concerned is completely persuaded that an overlapping code is impossible, but it is nice to have a foolproof general proof of it.[14]

In the formally published paper, Brenner added a concluding paragraph not included in the original Tie Club piece, which bears comment.

> This result [that all overlapping triplet codes are impossible] has one important physical implication. The original formulation of overlapping codes was based on the similarity of the internucleotide distance in DNA to the spacing between amino acid residues in an extended polypeptide chain. It was supposed that each amino acid was spatially related in a one-to-one way with each nucleotide on a nucleic acid template. The present result shows that this cannot be so and that each amino acid is stereochemically related to at least two, if not three nucleotides, depending on whether coding is partially overlapping or nonoverlapping. The difficulties raised by this can easily be overcome by assuming that the polypeptide sequence is in contact with the nucleic acid template only at the growing point, and detailed schemes can be readily proposed.[15]

Note that Brenner twice used the phrase "nucleic acid template," hedging his bets as to whether this was DNA (as suggested by Gamow) or RNA. His discovery of messenger RNA was still five years away. It is also relevant to point out that Brenner's suggestion that the genetic code may be founded on triplets of nucleotides, an assumption then quite widely held, was not yet formally proven. That contribution, too, lay in Brenner's (and Crick's) future.

Like many before him Brenner was immediately taken with Gamow. Brenner recognized a personification of himself in the former Russian physicist and can regale audiences with countless stories of Gamow's mischief. He particularly enjoyed rendering Gamow's broad Russian accent. "He had a strong Russian accent and you had to listen very carefully to understand him. Even worse was that he would use the words cytosine [the base in DNA] and cysteine [the amino acid] interchangeably." Some have said that whenever Gamow lectured at Woods Hole, a crew-cut gentleman in a dark suit would stand at the back of the room because the government had heard about this secret code that people were trying to crack—and it was very curious. " 'Cracking the code' was very much an in-phrase at Woods Hole during that summer."

The summer of 1954 at Woods Hole is also remembered by some for a grand party hosted by Gamow, but without his prior knowledge. This story, recounted by Watson in his memoir *Genes, Girls and Gamow*, was one of the times (perhaps the only time) that Watson out-pranked Gamow the prankster. Watson and Albert Szent Gyorgyi's son Andrew distributed invitations to a party to the entire scientific community at Woods Hole—in Gamow's name. For a while the master prankster was genuinely mystified by the many thankful responses he received. Once the plan was revealed, he graciously supplied the whiskey (his extreme fondness for which was widely known) and all at the Szent Gyorgyi residence had a fine time.

His brief stopover at the Cold Spring Harbor Laboratory in the summer of 1954 afforded Brenner the opportunity to cultivate relationships with several pioneers of molecular biology. Among them was the late Seymour Benzer, then a young assistant professor at Purdue University. The two had much in common. Like Brenner's father and mother, Benzer's parents fled Eastern Europe in the early 20th century. He grew up in New York and attended Brooklyn College before entering graduate school at Purdue to study physics, obtaining his Ph.D. in 1947 and an immediate faculty appointment in the

Department of Physics. One of the several physicists (and other scientists) who professed that their passion for understanding genes was ignited by reading Schrödinger's *What is Life?*, Benzer attended Luria and Delbrück's summer phage course at Cold Spring Harbor in 1948. "Three weeks of that," Benzer offered, "and I was converted."[16]

Benzer honed his bacteriophage skills with postdoctoral stints at the Oak Ridge National Laboratory, Caltech, and the Pasteur Institute in Paris. On his return to Purdue, he initiated experiments to map the genes of a bacteriophage, borrowing from a strategy invented by Martha Chase and Al Hershey (mentioned in Chapter 6). Essentially, it was known that when a phage particle attached to a bacterium its DNA entered the cell while its protein coat remained bound on the bacterial surface. Chase and Hershey showed that violent agitation (using a Waring blender) removed empty viral protein shells from the bacterial surface, but had no effect on the passage of DNA into the bacterium. These famous "Waring blender" experiments provided compelling evidence that the DNA, not the phage proteins, carries genetic information.

Benzer realized that interrupting the injection process at various times by using the Waring blender technique would capture pieces of DNA of varying size, thus offering the possibility of determining the sequence of genes on the phage chromosome. To achieve this goal he required a mutant gene as a reference marker "so [that] I'd know when a new gene would come through."[17] He selected the *rII* gene of a bacteriophage T4. Bacteriophage *r* (for rapid lysis) mutants are so named because they inhibit lysis of a particular strain of *Escherichia coli*, yielding morphologically distinct plaques on agar plates. (Plaques are clear zones of growth on agar plates caused by newly-generated phage lysing bacterial cells.) While optimizing these experiments, Benzer noted that whereas his *r* phage mutants readily yielded plaques on one strain of *E. coli*, they failed to yield plaques on a particular different strain.

In what he described as a "eureka moment," Benzer euphorically recognized that he had in his hands a genetic system that would allow highly refined genetic mapping of mutations in the *rII* gene. He could cross any two *rII* mutants and plate the progeny on the two *E. coli* strains. Mutant phages would produce no plaques. But if in any of the progeny there was a recombination event—a crossing-over between the two different *rII* mutants such that a wild-type recombinant phage free of both mutations was generated, it would yield a plaque.[18] Such was the sensitivity of this strategy that Benzer quickly understood that the simple trick of growing *rII* mutants on two particular strains of *E. coli* would allow him to identify mutations within the *rII* gene at single nucleotide resolution. Thus, he abandoned his original idea of crudely mapping phage genes and devoted his attention to a fine structure analysis of the *rII* gene of bacteriophage T4. This analysis was to revolutionize conceptual definitions of genes as functional entities, securing Benzer's place in the pantheon of biology.

Benzer arrived at Cold Spring Harbor at the end of June 1954 for a month-long stay. There is no record of his performing experiments there. Instead, he wrote a paper about the phage *rII* mapping project and its prospects, attended the frequent seminars offered during the summer, and engaged in discussions of his work on the fine structure of genes. Among the more fruitful of these conversations were several with Brenner. When Brenner met Benzer at Cold Spring Harbor in the summer of 1954, "he [Benzer] was carrying around a map of phage *rII* mutants and I was carrying around a book full of peptide sequences." As stated in the previous chapter, Brenner realized that if one could directly align base changes in DNA with corresponding amino acid changes in a polypeptide chain, one might be able to show that particular amino acids were coded by particular bases in DNA. Indeed, if one pursued such experiments to their limits, one may even be able to crack the genetic code. Thus the "colinearity problem," as this challenge was henceforth referred to, was hatched—and articulated in a paper Brenner published in 1959.

> Central to all modern theories of gene action is the hypothesis that the nucleic acid of a gene contains a coded sequence of nucleotide bases which specifies the amino acid sequence of the protein determined by that gene. Apart from the demonstration by [Vernon] Ingram that mutations can produce single amino acid changes in a protein, there is very little direct evidence for this hypothesis. Nucleic acids and proteins are both one-dimensional arrays and it is easy to see that a simple congruence might exist between the two structures. From the work of Benzer on the fine structure of the *rII* gene in T4 bacteriophage, it became clear that the relation between the linearity of a gene and that of a polypeptide sequence could be attacked experimentally. The simple question could be asked whether or not the order of mutations in a fine structure map corresponded to the order of amino acids suffering changes in the polypeptide chain. Since the map is some image of the nucleotide sequence in the deoxyribonucleic acid of the gene, this kind of study could show whether or not order was invariant when the gene transferred its information to protein.[19]

While at Cold Spring Harbor Benzer presented a seminar on the progress of his research with the phage *rII* gene. Shortly after completing a draft manuscript (which he distributed for comment to a few selected critics, including

Delbrück), Benzer departed for Europe. With Benzer's permission, Brenner, to whom he had also given a copy of his manuscript, read Benzer's paper at the annual phage meeting (Aug. 25 – Aug. 27, 1954). A few weeks later Brenner wrote to Benzer: "I presented your paper on your behalf at the phage meeting and must say it was favorably received. There were basically no arguments against it."[20] At Al Hershey's insistence Brenner prepared a summary of Benzer's paper for inclusion in a phage bulletin called the *Phage Information Service* that Delbrück distributed during the phage course.[21]

When he arrived at Cold Spring Harbor, Brenner was disappointed to discover that Watson had already departed for Woods Hole. He penned a quick note. "I arrived here a few days ago and have just heard that you are at Woods Hole. I intend coming up to Woods Hole some time but would like to time it so that I could see you."[22]

It has been said that Woods Hole, Massachusetts, "brings the Nobel Laureate and the local fisherman together in harmony."[23] Like Cold Spring Harbor, The Marine Biological Laboratory at Woods Hole is seductively appealing in the summer months. To this day the place is another recognized Mecca for scientists attending courses and formal meetings; intermittently sailing, walking on the beach, and partying late into the summer nights—all in the midst of intense intellectual exchange. Watson responded to Brenner's note: "I shall be in Woods Hole till the time of the phage meeting (Aug. 25) when I shall drive down to CSH. I would be delighted to see you here. . . . Francis Crick will also be here, as will Gamow. Why not come for a week or so?"[24]

Brenner cheerfully accepted Watson's invitation. In his reply he excitedly related the electrifying promise of Benzer's *rII* gene mapping studies.

> I have been having many discussions with Seymour Benzer who has some very interesting material on phage genetics. He has realized the experimental possibility of mapping the fine structure of a gene and has considered the results in terms of your model. I won't go into the details now, but I have a copy of his manuscript and an agency to discuss the matter thoroughly with you and Francis.[25]

It was at Woods Hole that Crick became fully appreciative of Brenner's penetrating intellect and his solid grasp of genetics. By the summer of 1954, Watson's hopes that deciphering the structure of RNA would yield further information about the genetic code were fading rapidly. Crick, too, had come to the sobering conclusion that cracking the code would require the strategic use of genetics and biochemistry, and he identified Brenner as "the ideal person to fill that role."[26] He went so far as to raise the possibility of a permanent position for Brenner at the Cavendish Laboratories after his mandatory return to South Africa. But nothing could be settled until Crick returned to England and successfully steered this plan past Max Perutz and the other senior scientists at the Cavendish. Still, Brenner was excited. Good fortune permitting, there seemed reason to anticipate his entrée to the famed Cavendish Laboratories in Cambridge to work side by side with Francis Crick.

Following this memorable, yet brief, interlude at Woods Hole, Brenner and Watson returned together to Cold Spring Harbor to attend the annual phage meeting, where Brenner presented Benzer's work on the phage T4 *rII* gene. At the meeting Brenner encountered another remarkable physicist who, like Gamow, made an enduring impression on him—Leo Szilard. Like Gamow, Szilard was a predictably quick learner and not at all shy about voicing opinions and criticisms at meetings outside his area of expertise. Brenner was also impressed when noting that Szilard, who sat in the front row for every presentation, would summarily rise and leave the room if he was not enjoying the talk. "He'd get up and walk to the door at the back of the room where he'd pause for a few seconds to give the speaker one final opportunity to please him. If the speaker didn't, as was usually the case, he'd leave the lecture room. Szilard was always scheming about one thing or another and we sometimes conjured up hilarious schemes together."

Immediately following the 1954 phage meeting at Cold Spring Harbor, Brenner was California-bound to visit institutions on the west coast, including Caltech, where he eagerly anticipated meeting the acknowledged father of molecular biology, Max Delbrück. He planned then to make his way north to the Virus Laboratory at the University of California at Berkeley, to work with Gunther Stent, a young phage geneticist. Watson was also heading west—to complete his somewhat desultory postdoctoral fellowship with Max Delbrück before taking up a faculty appointment at Harvard. When Watson extended an invitation to Brenner to join him in a car trip across the United States, Brenner delightedly accepted.

The pair first spent a night in New Haven visiting with one of Watson's uncles, a professor of physics at Yale and then headed for Boston to visit Watson's intended Harvard laboratory. Unknown to the two intrepid travelers, a hurricane (the infamous hurricane Caroline) had struck the East Coast. The couple arrived in Boston in an appalling rainstorm—but safe and sound.

The following morning they visited Watson's laboratory at Harvard and learned with shock (and relief) that they had been driving around Cambridge through the eye of the storm.

After a brief visit with Watson's parents in Chicago and a stop in Illinois to meet Salvador Luria, the pair headed west to California. While passing through a small town in Kansas, Brenner noticed a policeman following them. Brenner, who was driving at the time, anxiously asked Watson, "What do we do?" "Just drive," Watson muttered. But as they approached the Kansas-Colorado border, the policeman made his move and stopped the car. "You went through a traffic light. Get out of the car," he ordered. "I might have done so," Brenner apologetically replied. "I don't think I did—but if I did I'm terribly sorry."

"This was a Friday," Brenner recalled, "and Jim kept saying to me through the side of his mouth, 'Don't argue with him because he'll put us in jail and there'll be no judge 'till Monday and we'll have to spend the entire week-end in jail!'" The policeman asked to see Brenner's driver's license and duly established that Watson owned the car. After carefully examining Watson's car registration and noting the name "James Dewey Watson," the policeman asked: "Are you Mr. Dewey?" "Jim didn't particularly like the name and he most certainly didn't like the policeman calling him Mr. Dewey," Brenner stated. "I could see him getting very het up, so I kept telling him through the side of my mouth, 'Don't talk back to him or we'll be in jail for the week-end!'"

Driving together for several days, Watson and Brenner had ample opportunity to discuss science, although there really was not that much to talk about at length. "Everything was plans then." At Caltech Brenner again presented Benzer's work on dissecting the phage *rII* gene. He had no reservations about announcing his and Benzer's hope that exhaustive mutational analysis of the *rII* gene, coupled with amino acid sequence analysis of phage *rII* protein, might demonstrate strict colinearity between nucleotides in DNA and amino acids in the protein. About a year later Benzer gave his version of the *rII* experiments at a series of west coast seminars. He wrote to Brenner, then in South Africa, jokingly informing him that he was somewhat put out to be told, "Frankly, Sydney tells the story better than you do!"[27]

Following a brief vacation lazing on the Southern California beaches and hiking the trails that weave through the coastal foothills, Brenner headed north to the San Francisco Bay area to join Gunther Stent, a phage and bacterial geneticist who contributed an early textbook on molecular biology. Like Delbrück, Stent had fled Nazi Germany as a youth and, in fact, became an early member of Delbrück's phage group before launching his own laboratory at the University of California at Berkeley. Aside from their mutual interest in bacteriophage and bacterial genetics, Brenner and Stent delighted in their shared sense of humor, perverse though it sometimes was. Their correspondence is replete with madcap salutations and countless puns and jokes—contributed by Stent with characteristic affectation. "Ta ever so for your Oxonian and Cantabrian letters which both Inga [Stent's wife] and your serviteur devoured with the greatest of that old gusto," Stent once penned.[28]

Brenner's extended stay in Berkeley presented ample opportunity for extended discussions with Stent and others. The many topics they talked about included the use of bacterial protoplasts[b] as sub-cellular systems for biochemical studies. Protoplasts are readily permeable to many biological molecules, thus offering the tempting possibility of being manipulated for intracellular studies. The pair was especially pleased to discover that they could infect protoplasts with bacteriophages, and when Brenner returned to South Africa, he continued these experiments. Unfortunately, the tantalizing notion of using bacterial protoplasts to dissect phage molecular biology never materialized.

Back in England to set sail for South Africa, Brenner visited briefly with Francis Crick and his wife Odile in Cambridge. Coding occupied much of their discussions, as did the paramount issue of securing a permanent position for Brenner at the Cavendish Laboratories. To that end Crick introduced Brenner to senior members of the Cavendish staff, including Max Perutz, Director of the Medical Research Council Unit for Molecular Biology at the Cavendish Laboratory. Before boarding a boat back to South Africa, Brenner wrote disconsolately to the Cricks: "I hope to see you all soon: but now I am furiously concentrating on leaving this sceptred isle. It is gray here in Southampton; I feel rather gray myself. It's all slipping away and I can do nothing about it."[29]

Brenner arrived in South Africa in late 1954 to assume an appointment as Lecturer in the Department of Physiology at his alma mater, the University of the Witwatersrand. In his obligatory report to the Carnegie Foundation of New York he wrote: "I can say without any doubt that I benefited immensely from my visit to the United States. As Lwoff[c] has put it recently in a review

---

[b] Protoplasts are viable bacteria in which the chitinous outer layer of the bacterial cell is removed to render them permeable to exogenous agents.

[c] Andre Lwoff was a famous French microbiologist and one of the pioneers of the study of bacteriophages.

article: 'The United States are now, par excellence, the holy land of bacterio-phage.' "[30]

The roughly three-year period abroad had widened Brenner's horizons beyond his expectations. His encounters with Watson and Crick and his fateful glimpse of the DNA structure had unequivocally focused his career goals. His future was no longer restricted to becoming South Africa's reigning "bright boy" in biology. There was a distinct probability that he would, in due course, be traveling again to England, this time to work with Francis Crick and collaborating with some of the rising young stars in the infant science of molecular biology. The future looked promising. Brenner stated:

> At the time that I left England I knew exactly what I wanted to do. I had formulated what had become for me the gene-protein problem. We were now going to investigate the correlation between genes and proteins by finding a gene on which we could do fine structure analysis and then its corresponding protein on which we could do chemical sequencing. The object was to prove colinearity; namely that the mutations occurred on the gene in the same order with which they were present in the protein. So on the way back to South Africa I spent a lot of time thinking about how exactly to tackle this colinearity problem.

8

Returning to South Africa

I didn't think that one could do any good science there

I MMEDIATELY UPON HIS RETURN TO SOUTH AFRICA, Brenner (grudgingly) assumed his duties as Lecturer in Physiology at Wits (a faculty position equivalent to that of assistant professor in the American university system). Joe Gillman, his erstwhile mentor as a science student and the newly appointed department head of the Department of Physiology, ambitiously sought to mount a state-of-the-art course for medical and dental students. He collected physiology curricula from many of the leading medical schools in the United States, imploring his small faculty to use these as guides for designing the course. Much of the organizational work fell to Brenner, who had to contend with limited departmental resources to mount any sort of respectable program.[1] But Gillman was adamant. "Joe could be very difficult," Brenner stated. "He had all sorts of ideas, but no talent for implementing any of them ... And he had to win every argument! He used to say allegorically: 'Take the liver'—meaning consider the liver as an example. But I would say to him; 'No Joe, you take the liver. I'll take the spleen.' When I first met Jonas Salk years later he reminded me a lot of Joe in his appearance and mannerisms. And when he said to me in the midst of a scientific conversation: 'Take the liver'—I almost said: 'No *you* take the liver!'"

Brenner presented didactic lectures in physiology to medical and dental students. No matter how shallow one's interest in physiology, students (this author included) rarely fell asleep for fear of missing one of his hilarious anecdotes. Rumor had it that he spent most of his time in a research laboratory, though no one knew quite what he was up to. What Brenner was in fact "up to" was bacteriophage genetics. He set up a functional laboratory in the Department of Physiology, obtained a small research grant from the South

African Institute for Medical Research, and hired a bright young technician, Betty Van Zyl.

During his two years of "mandatory confinement" in South Africa, Brenner maintained a robust correspondence with Crick, Stent, Benzer, and Watson. Letters from abroad kept him in touch with scientific progress in the "real world" and raised his spirits when they flagged—not an infrequent event when he first returned. In the final days of 1954, he optimistically communicated with Gunther Stent:

> It was so nice to arrive here and find a letter from you waiting for me. Let me tell you what my setup is here. I have a job as a Lecturer and will be promoted to Senior Lecturer next year. The salary isn't bad; I get 1400 pounds a year. . . . I have a lab about twice the size of yours to myself and am busy getting together pipettes, tubes, etc. I have the tanks (baths) and incubators coming and in a month should be well set. I am trying to get money for isotope equipment and for fellowships for graduate students—this should be easy. All in all, I am quite pleased. Furthermore nobody knows what I am doing so I am not bothered by long discussions about nothing.[2]

Within a short time, however, signs of frustration and isolation in South Africa emerged. On January 5, 1955, Brenner wrote to Watson:

> 5 January 1955
>
> Dear Jim,
>
> I have at last settled down here and hope to start working soon. The problem of collecting apparatus has assumed quite large dimensions but it should be solved in the next few weeks. . . . Do write when you have time. I feel so cut off and isolated here that a note from you telling me about your work would be most welcome.[3]

"I was extremely depressed at times," Brenner stated. "I sometimes thought that I'd never see America again. I even sometimes had the feeling that coming home was sort of the end of everything—just when I was getting started. But I kept reminding myself that the wheels were turning to return me to England." In April 1955 he again wrote plaintively to Watson: "I work like a slave and all the difficulties are increased by absolute isolation. So you see, I can sympathise with you when you say that you have no one at Caltech to talk to. I feel that I must get away next year to keep in touch, but we are so far away and money is hard to get."[4]

Crick was determined to go to any lengths to bring his newfound South African colleague to Cambridge. In early 1955 he wrote encouragingly: "Sorry to hear about your difficulties. However, everyone is agreed that we must get you back here, and it is just a matter of waiting till you think the time is ripe, so don't take them to heart."[5]

But others in the United Kingdom and America had recognized Brenner's intellectual powers, his passion for science, and his boundless energy, and had their own designs on dislodging him from South Africa. A letter from Gunther Stent, written in early December 1954, stated: "Boy do we miss you here! You simply <u>must</u> come back to Berkeley. In future letters I hope we can discuss in more detail just what we can do to bring you and your gang for an extended stay at Baghdad-by-the Bay."[6]

Brenner gravely replied: "All of my American trip is slipping by and it seems like a dream, and I desperately try to maintain the tenuous strands by writing to my friends. [But] at the present moment, I do not think of going away again. I feel I must settle down and do a solid piece of phage work. Perhaps in a year or so I shall think again."[7]

Brenner also received a letter from Josh Lederberg,[a] offering him a faculty position at the University Wisconsin. Wondering how Lederberg had even entered the picture, Brenner subsequently discovered that Seymour Benzer had included his (Brenner's) name as his co-investigator on a research grant proposal. The proposal was seen by Lederberg (presumably he reviewed it) who decided that if Benzer was trying to recruit Brenner he would too. Perhaps spurred on by this news, Benzer made an unambiguous offer to Brenner to join him in Purdue. The Department of Zoology at Edinburgh University had not given up on him either. But having visited the place in mid-1955 and discovering the serious intention to recruit Brenner to Scotland, Crick did not hesitate to remind his young colleague of the possibility of a position at the Cavendish. In early July 1955 he wrote to Brenner again.

> I think the time is approaching when we should give serious thought to you coming back to England. I think you will have no difficulty in going to Edinburgh, but naturally I should much prefer to see you here. ... If there is a possibility of arranging matters to start a year from now we ought to think about it fairly soon.[8]

---

[a] Joshua (Josh) Lederberg, a pioneer in bacterial genetics, greatly enriched the early revolution in molecular biology. He served as professor of genetics at the University of Wisconsin, then at Stanford School of Medicine, before coming to the Rockefeller in 1978. His life-long research, for which he received the Nobel Prize in 1958 (at the age of 33), was in microbial genetics.

A few months later Brenner received another encouraging letter from Crick:

> There is some prospect of doing something by next academic year (i.e., Oct. 1956). Could you please send me, as soon as possible, the full details of your education, research experience, papers published, etc. Also the salary you would like and the <u>minimum</u> salary you will accept (this last figure in confidence to me) bearing in mind that once you are here you should easily be able to increase your salary.[9]

Despite these reassuring sentiments, Brenner's concern about the possibility of Crick's efforts coming to naught prompted him to maintain several other possibilities, including the offer to join Gunther Stent as a staff member in his Berkeley laboratory. Aside from his unequivocal preference for joining Crick at the Cavendish Laboratories, thoughts of relocating anywhere in the United States were always tempered by concerns about visa problems for May, as well as her general ambivalence about the country. Adding to the strain of the uncertainty of his future, in October 1955 Brenner received news from Crick that was by no means encouraging.

However, about a month later he received yet another communication from his enthusiastic supporter that gave him cause for muted celebration.

> My Dear Sydney
>
> I hope you will not think, from the rapid change between my last two letters, that I am constitutionally unstable in my opinion. The two main reasons for the change were:
>
> 1. We heard from the Rockefeller that they were going to give us $40,000, spread over 4 years. This changes a position of acute financial stringency into one of comparative plenty, since it supplements our MRC [Medical Research Council] allowance.
>
> 2. A senior member of the unit decided to leave, thus making it possible for us to put up to the MRC a case for someone in his place.
>
> I therefore seized the opportunity to put a strong case that you should come here and be loosely attached to me. This case has much force, because I have not previously asked for anyone to work with me, and thus my first request was bound to be sympathetically received. Max Perutz and John Kendrew are now agreed that if we were to arrange it you should join the unit. The major difficulty is one of space, which still remains acute. I think eventually there will be extra space in the Cavendish, but we are unlikely to have it by next year. I am therefore exploring the possibility of finding you temporary space (plus a desk in the Cavendish) . . . in the Molteno, or with Michael Stoker in Pathology. It is of course, understood that whatever the arrangement you will be completely free to work on what you please. Could

you let me know how you feel about this sort of thing; also some idea of the problems you might want to tackle and the facilities you will require. I think the Rockefeller money will cover any reasonable demands for apparatus.

Yours ever,
Francis[10]

In his response, Brenner volunteered to work in a cupboard if necessary. To which Crick blithely replied: "I was delighted to hear that you were prepared to work in a cupboard! However, in case there is only half a cupboard I have been exploring other possibilities."[11]

On December 30, 1955, Crick sent Brenner a belated Christmas present by way of a letter indicating that Harold Himsworth, Chairman of the MRC, "has agreed that I may 'open informal negotiations' with you for 'an appointment of limited tenure' (which means 3 to 5 years) on the Council scientific staff—starting January 1957. I need hardly say that I should advise you to accept."[12]

Brenner accepted, of course, but not before he and May (who was less insistent about leaving South Africa) examined their futures in a series of protracted discussions. Among other issues that merited consideration was the sobering realization that the Cambridge offer was for a mere eleven hundred pounds a year (equivalent then to about US$2500). With a third child recently born (a daughter, Belinda), the couple knew that it was going to be challenging to manage financially.

Aside from such pragmatic considerations the couple, May in particular, questioned the moral aspects of the decision. Should they remain in South Africa and fight the good fight, or should they leave, like so many other liberal South Africans were doing, perhaps carrying the guilt of having deserted their homeland? In the end, May conceded that her husband's career goals were simply too overwhelming to be denied. "I remember him telling me in the most sincere terms that he was interested in many things—but that he sincerely believed that his purpose in life, so-to-speak, was to be a scientist. And while he knew full well that he could end up as a big fish in South African science, the pond there was small and he wanted to know whether he could make it as a big fish in a big pond. So we left."[13]

Brenner never seriously entertained the idea of remaining in South Africa permanently. "I didn't think that one could do any good science there. The proving ground for me lay in the international science community and I knew that I had to test myself there to know if I really was any good as a scientist." Additionally, he viewed South Africa's future bleakly. Like many liberal white South Africans, Brenner was convinced that if the African majority in South Africa ultimately won politically, a bloody civil war would ensue. He was concerned that such a calamity might push the country to the same depressing fate that other African countries had suffered. "So May and I made arrangements to move our three children and ourselves to Cambridge, England. I was enormously excited about returning there, especially since Seymour Benzer and I had arranged to pursue the colinearity problem together there."

Events moved swiftly and without undue bureaucratic hassle, prompting Max Perutz to later comment: "No panel, no referees, no interview, no lengthy report, just a few men with good judgement at the top."[14] Crick remained in close contact with Brenner, imparting advice and council in all manner of practical matters, such as schooling for the Brenner brood, securing a relocation allowance from Perutz, and finding suitable housing. "Crick's bustling cheer never varied; his attentive kindness was unflagging."[11] The two men were to share an office for 20 years.

Meanwhile, Brenner diligently pursued his duties as a lecturer in physiology, and in his ample spare time, he succeeded in identifying phage mutants defective in a cofactor required for phage adsorption to bacteria, in the hope that this might prove useful as an alternative to phage rII protein for the anticipated colinearity studies with Benzer. He published these results,[15] but like the intractable phage rII protein, phage adsorption factor was not destined to occupy a place in solving the so-called colinearity problem.

In March of 1956, Brenner formally notified the University of the Witwatersrand of his intention to resign his appointment as lecturer in the Department of Physiology, effective December 31, 1956. His neatly hand-written report on his research supported by the National Cancer Association of South Africa provides a succinct summary of his efforts during the preceding two years.

> A. Experiments were continued with protoplasts of *Bacillus megaterium* in an attempt to demonstrate the direct roles of nucleic acids in bacteriophage formation. These experiments were uniformly disappointing. . . .
>
> B. Investigations have been made of the early stages of bacteriophage T2 synthesis using radiogenetic techniques. . . .
>
> C. Work was begun on the problem of determining the structure of the tail proteins of bacteriophage T2. As a preliminary, a study of the genetic control and phenotypic characteristics of the absorption cofactor was carried out. . . .

D. Theoretical work on the coding of amino acid sequences by nucleic acids has been carried on. A summary of the work has been privately circulated to the group interested in this work and may be published later.[16]

In the waning days of 1956, Brenner, together with May and their three children, docked at Southampton to begin his three-year appointment at the MRC Cavendish Laboratories. May found work as an educational psychologist with the Cambridge Town Council and, in later years, opened a private practice in psychological counseling. Parenting consumed much of her time and attention. Following Belinda's birth another daughter (Carla) enlarged the family.

Though barely on the doorstep of his career, any lingering question in Brenner's mind as to whether he was "really good or not" had long since evaporated. Having already been tested among the intellectual elite in modern biology, he was not found wanting. He would never again be conflicted by considerations of a permanent return to the country of his birth. Nonetheless, South Africa never forgot its favorite scientific son. He has been coaxed back for brief visits to celebrate many of his professional successes and has been feted with all manner of honors and awards, including the Order of Mapungubwe (Gold),[b] the country's highest honor and one personally granted by the President of South Africa. Mindful of his modest beginnings, Brenner in turn established a postdoctoral fellowship in his name, jointly administered by the Academy of Sciences of South Africa and the U.S. National Academy of Sciences.

---

[b] The Order of Mapungubwe comes in four categories: platinum, for exceptional and unique achievements; gold, for exceptional achievements; silver for excellent achievements; and bronze, for outstanding achievements. Nelson Mandela is the only recipient in the platinum category.

PART 3

Deciphering the Genetic Code

9

Cambridge at Last

We were completely lost, you see

Sᴍᴍ

Single-mindedly burying himself in work was nothing new for Brenner—or his family. As a working professional for much of her life, May had long adjusted to the reality of her husband's passion for science. "You don't get to be a great creative scientist when you are also an attentive husband and father," she once stated. As much as she and their children wished they could have more of Sydney's time, they accepted the reality that there were not enough hours in the day for the man to pursue his commitment to science and to be a constantly attentive husband and father. "The children both accepted and resented the reality that this was what Sydney was all about—and still is."[1] Nor is Brenner oblivious to his family's tacit support. When providing autobiographical information to the Nobel Foundation he wrote: "My wife and family have borne the burden of a preoccupied husband and father for 50 years. Living most of the time in a world created mostly in one's head does not make for an easy passage in the real world."[2]

But occasions when Brenner was at home (and mentally and emotionally accessible) were typically filled with his boundless sense of humor and playfulness, which naturally endeared him to his younger children. "He was always telling the kids silly jokes and expressing his humor," May related. "Once one of our grandchildren asked me: 'Grandma, did grandpa have to learn to be so silly, or was he always like that?' I told her that she was asking the wrong grandparent! So she asked Sydney the same question, and he replied: 'Oh yes. I had to learn very hard to be so funny.'"[1]

Jonathan, now a man in his middle age who has devoted much of his life to humanitarian pursuits in war-torn parts of the world, never confused Brenner's and his biological father's roles in his life. When Jonathan was a young child in South Africa, Brenner never interfered with the boy's relationship with his biological father and rarely exercised paternal discipline unilaterally. "In later years Sydney and I were in parallel worlds with separate relationships with my mother," Jonathan stated. "But we were and are good and respectful friends."[3] Brenner would occasionally take his stepson to his laboratory, and to Jonathan's delight once showed him how to blow glass to fashion pipettes. He also devoted considerable time and effort to helping his stepson with scholastic studies. "I would ask him things as a school child and he would take enormous amounts of time to explain things to me," Jonathan continued. "He taught me the importance of clarity in my writing, telling me that if someone cannot communicate what they want to tell you about in the first sentence they write, the rest is probably not worth reading. I've never forgotten that as an object lesson whenever I write."[3] Although not especially interested in science, Jonathan cultivated a deep respect for his stepfather's scientific reputation. "I came to the realization that there is nothing in biology that Sydney doesn't appreciate or understand—as I tell people when I show them his Nobel Prize medal."[3]

Brenner's youngest daughter, Carla, an attractive woman with a family of her own, was also blessed with a burning passion from early childhood—for ballet, a career choice that was tragically aborted after she was hit by a car, injuring her pelvis. She now teaches Pilates in a studio at her London home. Carla relates fond memories of growing up around her father. She, too, was treated to visits to the laboratory on occasional weekends.

> I remember counting things [presumably bacteria or phage on agar plates]. He'd say: "OK Carly, 1, 2, 3, 4, 5"—and move on to the next one. [referring to the common habit of counting batches of five]. He would have me sit next to him, pretending he needed my help with things.
>
> I loved to be with my Dad because he was always so funny. I can't remember exactly when I began to appreciate his humor, but it was when I was very young. He'd do funny voices and we had a made-up language of our very own. And he would sing to me when I was very little and read stories to me. It was a magical time for me. He wasn't around very often, so it was always special when he was, particularly when he focused on me. My siblings and I knew that there was this mythical place called the lab. And we knew from an early age that that's where he went a lot of the time, and that's where his passion was. But mainly what I remember as a child is that when he was around it was magical.[4]

Married to an accomplished musician and having studied ballet seriously, Carla is no stranger to passion in one's work.

Brenner lost no time mounting a research program. Primarily designed for structural biology, the Cavendish had limited resources for phage experiments, and Brenner once again faced the ordeal of assembling a functional biology laboratory essentially from scratch. Crick had warned him that space was tight and, indeed, it was—so tight that for a time he and Crick shared an office with as many as five other scientists. At one time the office also housed a dishwasher that doubled as an autoclave.

Benzer and Brenner maintained frequent contact about progress on the phage *rII* gene-mapping story. In the spring of 1956, Benzer wrote: "There are now so many mutants on the map that one can start looking at the thing through protein colored glasses, but no great truths have emerged from this . . ."[5]

As we saw in Chapter 7, if Benzer was able to resolve phage mutations in the *rII* gene at the level of single nucleotides, he very likely could demonstrate that certain nucleotide substitutions led to corresponding amino acid substitutions. Not surprisingly, others were pursuing this goal, especially when it became apparent that purifying suitable quantities of phage rII protein was going to be enormously challenging. Benzer was somewhat dismayed to discover that George Streisinger, a young microbiologist who (like Jim Watson) had learned phage genetics from Salvador Luria, was doing the same experiments but with a different gene-polypeptide combination.[6] In bacteriophage T2, a gene called *h* (for *h*ost range) controls the absorption properties of phage particles to different bacterial hosts through the expression of phage tail fibers. Thus, for example, phage carrying a wild-type *h* locus can absorb to a strain of *Escherichia coli* called strain B, but not to a different strain called B/2. In the mid-1950s Streisinger and his colleague Naomi Franklin began studying the genetic properties of the host range region of phage T2 and determined that they could map host range mutants in much the same manner that Benzer was doing with the *rII* region.

Benzer's concerns were heightened when he received a letter from Streisinger in early December 1954:

> We really have been overwhelmed with your experiments—the idea is really beautiful. What is the latest poop? How close can you get two markers? Inspired by you I have been thinking of starting some experiments along the same lines with the *h* locus—the whole thing will depend on whether the locus is complex. The beauty here would be that the protein is accessible.[7]

Benzer realized the extent of Streisinger's interest in the colinearity problem when he attended a small symposium arranged by Max Delbrück at Caltech in March 1955. Streisinger was not present, but his coworker Naomi Franklin discussed her experiments on fine structure mapping of *h* mutants of phage T2. Benzer took careful notes and later wrote to Brenner and others that Franklin and Streisinger had observed a cluster of mutations that looked very similar to what he had discovered with the *rII* gene. Benzer was impressed, but not entirely pleased. In short, he was now "feeling the hot breath of competition."[7] To be sure, shortly before arriving in Cambridge, Benzer attended a summer Gordon Research Conference where he learned that at least four different groups were working on phage proteins and fine-scale genetics to address colinearity and the coding problem.[8]

Streisinger also encountered problems in purifying sufficient protein for colinearity studies and became increasingly skeptical about the utility of the host range system. In mid-1956 he wrote to Brenner, who was then still languishing in South Africa.

> The purpose of all this originally was to have a system in which the genetic analysis and an analysis of the product were both feasible. I wonder however whether phage is really the best thing to use this for! It is probably quite difficult to get out the tail tip protein in large enough quantity to be useful, to separate it from other possible tail proteins, etc. . . . I myself am not planning to go ahead with this, but Naomi Franklin, who did the genetic analysis with me _is_ going to try to separate and purify a tail protein and do analysis. If you would want to do anything with the *h* system I would be glad to furnish you with strains and information etc.[9]

A few weeks later Streisinger again wrote to Brenner, this time expressing his interest in tackling this daunting problem in collaboration with Brenner when he moved to Cambridge.

> [I am seriously thinking of] going to Cambridge for a year or two to learn about structure and perhaps try my hand at the phage tail protein. This seems especially appealing since you'll be there: perhaps we could do part of this together. . . . This shouldn't of course limit anything you want to do in the meanwhile; as I had written to you earlier I would be glad to send you any strains (various *h* mutants, etc) that you may want to have, and you could proceed to do whatever you liked with these. The problem certainly won't be solved within two years, so that there will still be things to do when I get there.[10]

Brenner was unconcerned about collaborating both with Benzer and Streisinger. His primary quest was to demonstrate colinearity between the nucleotide sequence of a gene and the amino acid sequence of the protein it encoded. Any latent concerns about stirring up a potential clash between Streisinger and Benzer were obviated in the summer of 1956. Brenner, still in South Africa, received a long letter from Crick that primarily addressed the topics of a recent meeting he attended in the United States, but also revealed his own indifference about such trifling matters.

> Streisinger reported the beginning of Benzer mapping in the tail attachment locus. Incidentally, he asked about coming to Cambridge later to work with you in looking for tail protein. Benzer is also trying to find the protein for his [rII] locus. I told both of them to come [to Cambridge] whenever they felt like it.[11]

Streisinger and Benzer joined the Cavendish Laboratory at about the same time, in the fall of 1957. Together with Brenner and Sewell Champe, a graduate student in Benzer's laboratory, they lent their collective weight to the colinearity problem. Prior to his arrival Streisinger again voiced his concern about the huge quantities of host range protein required to carry out the peptide fingerprinting techniques recently worked out by Vernon Ingram[a] (also at Cambridge). But the resourceful Brenner hit on the notion of adapting a Hoover washing machine to grow huge quantities of phage-infected bacteria and wrote to the Hoover company. "They actually gave me a machine for free that looked as if it would work perfectly. But after the first run the aluminium lining was so corroded by the growth medium we had to throw the thing away." The colinearity experiments remained stuck.

Meanwhile, Brenner shifted his studies of phage tail proteins to the emerging field of electron microscopy. "I used the electron microscope to look at phage and began doing experiments to see if I could take phages apart and identify different structures (like tail fibers) in the microscope." He discovered that if he simply left phage in a solution at a very acid pH and then digested this with various enzymes, he could observe intact phage tail sheaths under the microscope. Applying his earlier experiences with negative staining learned from Alf Oettle in South Africa, Brenner and his collaborator Robert Horne rendered stunning electron microscopic images of phage particles in

---

[a] By the summer of 1956, Vernon Ingram, an English chemist who can justly be called the father of molecular medicine, had shown that a single amino acid substitution was the sole abnormality in hemoglobin from individuals with hereditary sickle cell anemia. Key to this success was Ingram's idea of digesting hemoglobin into shorter segments using trypsin, which yielded a limited number of peptides that were amenable to analysis. This technique was called protein or peptide fingerprinting.

various states of disassembly. "I wasn't interested in pursuing this in depth. But this technique gave virologists a great tool and allowed people to think about investigating macromolecular assemblies." Brenner and Horne published these studies in 1959.[12]

Benzer arrived in Cambridge in October 1957 and, as one who detested cold weather, had a difficult time adjusting to the British winter. Brenner was concerned enough that he would sometimes go to Benzer's flat and light fires for him. Keeping warm was a constant preoccupation, prompting Benzer to purchase a thick sweater of the type used by fisherman on the Faeroe Islands, which he was rarely seen without. When in the poorly heated Cavendish Laboratory, he hovered around the small heater under technician Leslie Barnett's desk.

Benzer was also a night owl who enjoyed working into the wee hours of the morning and sleeping late. One morning shortly after the 1958 Nobel Prize in Medicine was announced (that year awarded to the geneticists George Beadle, Edward Tatum, and Josh Lederberg), Brenner received a phone call from a *TIME* magazine reporter, asking to speak to Dr. Benzer. When Brenner informed the journalist, who was presumably hoping for a comment on the Nobel Prizes, that Benzer was asleep, the journalist asked whether he might get hold of Dr. Archibald Garrod.[b] Brenner asked the caller, who was obviously unaware that Garrod was long since dead, to hold the line while he went to find Garrod's address. He quickly looked up Garrod's obituary and established that the man was buried at the Highgate Cemetery. "So I told them this, and also informed them that they stood a better chance of waking him up than of waking Seymour!" Benzer, who died in late 2007, maintained this odd working schedule to the end of his life, arriving at his laboratory in mid-afternoon, returning home to eat dinner with his wife before seeing out a full night of work in his Caltech laboratory.

Whatever else about which Brenner, Benzer, and Streisinger might have disagreed, the trio emphatically concurred that English food in the late 1950s was simply awful. Prior to his arrival, Benzer had diligently researched Cambridge restaurants, some of which looked very promising—at least on paper. As soon as he joined the laboratory, he eagerly suggested to Crick that they try a different restaurant for lunch each day. However, this promise faded rapidly when put to the test. "One had a choice between a bad Chinese

---

[b] Archibald Garrod, whom many considered the founding father of molecular genetics, was the first person to relate a human disorder with Mendel's laws of inheritance through his work on alkaptonuria. He also proposed the notion of inborn errors of metabolism. Garrod died in 1936.

lunch—and a terrible Chinese lunch," Brenner was heard to say. When George Streisinger arrived about a month later, his enthusiastic suggestion that the group explore the Cambridge restaurant scene was met with gales of laughter.[13]

As an alternative to their frustrations in the local eateries, Benzer, Streisinger, and their wives, together with Renato Dulbecco (another Luria-trained molecular biologist and a future Nobel Laureate) and his wife initiated a gourmet dinner club devoted to sampling English game, including grouse, partridge, hare, guinea hen, and fish.[13] In due course Brenner, Benzer, and Streisinger organized a lunch cooking club as well. Streisinger once traveled all the way to Yarmouth to procure fresh mussels to prepare *moules mariniere*. Eventually things got to the point where the scientists were spending more time preparing gourmet meals than in the lab. However, it all ended one day when it was Benzer's turn to make lunch, and he walked in with packages of fish and chips, sheepishly apologizing for the fact that he woke up too late to prepare a proper meal.

Brenner and Benzer remained close friends and scientific colleagues for many years and shared several prominent scientific awards. "There was a strong cultural affinity between us, and we shared a common view of the world," Brenner commented. The pair shared a particular passion for pickled herring, references to which occur in much of their extended correspondence. Writing about a seminar that Brenner was scheduled to give at Caltech in late 1999, Benzer stated: "We will be delighted to hear a talk about Fugu. But unless your Japanese connections can supply some, we will have to make do with herrings." About three years later Benzer observed, in an e-mail message to Brenner, "In case you are not aware of it, the ultimate herring is to be found in Sweden. It is called *surstromming* and is quite an experience."[14]

The intractability of the phage rII and h proteins continued to frustrate the Cambridge scientists, leading Brenner to the conclusion that "a tractable system is crucial because if the question you're asking is general enough it can be extrapolated to other biological systems. I believe that researching experimentally tractable systems to ask and answer specific questions is a good way to spend one's time." Brenner imagined that somewhere there must exist an ideal organism for colinearity studies. It would be endowed with 28% of all its protein in a single polypeptide, and the protein could be crystallized by simply bubbling carbon dioxide through an extract. Of course this dream bacterium would also be perfectly suited for high-resolution genetics. "Sometimes we used to jokingly say: 'Well, maybe we should look for this perfect bacterium.' " On a more serious note, when Brenner once encountered a paper that described the impressive inducibility of cytochrome C when certain strains of *Pseudomonas* are grown under specific conditions, he managed to persuade a young man in Fred Sanger's laboratory to purify the protein. "But there was simply no genetics in *Pseudomonas,* so we dropped the whole idea."

By the fall of 1957, a number of laboratories around the world were attempting to demonstrate colinearity of the gene and its polypeptide product, but to no avail. Colinearity proved intractable to everyone. In a 1967 letter to Benzer discussing a lecture that he was planning to present at a symposium, Brenner wrote: "At the back of my mind was the idea that I should really talk on the rise and decline of the *rII* gene, but Francis vetoed that and I shall be giving a straight talk and only hinting at the joy and sadness."[15] The fundamental tenet of the colinearity experiment was sufficiently compelling that enterprising investigators continued to visit the issue. Nonetheless, it was not until 1964 that Charles Yanofsky and his colleagues at Stanford University succeeded by investigating the *E. coli* tryptophan synthase gene and protein. "Yanofsky started from the other end, if you will," Brenner commented. "He had worked on tryptophan biosynthesis in bacteria since he was a graduate student and he was smart enough to realize that the tryptophan synthase gene and protein were experimentally well suited for the colinearity experiment." In a rare tribute to the discipline of biochemistry, Brenner wrote in a review article:

> [The] idea that one could do everything without opening the black box of biochemistry between genes and proteins, although born out of desperation, soon acquired a fascination of its own; it was immensely attractive and generated much of the style of the subject. But, of course, as everybody knows, the programme of molecular genetics was never completed. The expectation, in 1954, that it would take a long time to understand the detailed biochemical machinery of information transfer and protein synthesis was hopelessly pessimistic.[16]

Despite the disappointing outcome of his research efforts during his stay in Cambridge, Benzer's sabbatical was, by no means, a failure. In the course of their many conversations, he and Brenner developed an enterprising notion that was a direct extension of the colinearity experiments. Brenner explained:

> We reasoned that if we could demonstrate that some mutagenic agents generated specific types of mutations, we might be able use this information for breaking the code. For example, if we had a chemical reagent that say always changed guanine to adenine, and a different agent that say always generated

the mutation of cytosine to thymine, and if we could correlate these mutations with changes in the amino acid sequence of the relevant protein, we should be able to decipher at least part of the code.

Benzer had, in fact, done elegant work with Ernest Freese at Purdue, demonstrating a provocative correlation between the types of mutations that arose in phage DNA and the type of mutagens to which they were exposed. Brenner was particularly struck by the observation that, whereas mutations induced by 5-bromouracil were revertible (i.e., at some measurable frequency nonmutant offspring could be derived), spontaneous mutations were not. "The fact that spontaneous mutations were non-revertible was an interesting puzzle to me." Brenner was to solve this puzzle when he pursued experiments to understand the phenomenon of nonsense mutations (described in Chapter 14).

One can trace Brenner's interest in mutagens and mutagenesis to his early preoccupation with dyes as a student in South Africa, an interest that was notably heightened by a 1953 report in the literature by Robert "Bob" DeMars (another member of the Salvador Luria stable of molecular biologists) that a dye called proflavine, a polycyclic aromatic compound with a planar configuration, was a potent mutagen. Brenner and Benzer, with the help of Leslie Barnett, an able technical assistant, began executing experiments aimed at understanding how chemical mutagens (including proflavine) work. For Benzer this was an obvious continuation of his extensive mutagenesis studies using the phage T4 *rII* gene. The group examined the mutational spectrum in T4 phage exposed to proflavine and compared it to the spectra generated spontaneously and following exposure to 5-bromouracil. They discovered that the latter compound also had a characteristic mutational signature. In a paper in *Nature* published in late 1958, they wrote, "The contrast between the three mutational spectra underlines the complexity of the genetic fine structure. A cistron (see below) evidently is composed of a large number of mutable elements of several kinds. It should be fascinating to relate these local details to the precise chemical constitution of the genetic structure. . . ."[17]

Though Benzer's fine scale dissection of the phage T4 *rII* gene never realized the hope that he and Brenner entertained with respect to colinearity, his work led to the important insight that a gene can be operationally defined in different ways. He not only coined the term "cistron" to define the gene as a functional unit, but also invented "recon" to define the gene as unit of recombination, and "muton" to define the gene as a mutational unit. "Cistron" became permanently embedded in the lexicon of biology, as did "codon," a term contributed by Brenner. "Recon" and "muton" became the subjects of a droll linguistic controversy and were soon dropped. According to Brenner, "recon" sounded too much like "reckon" to American ears, and "muton" faced stern objections from François Jacob, who, after hearing the word mentioned in a seminar, quizzically asked "what all the stuff about sheep meant."[18]

Not too much later, after it was demonstrated by others that proflavine can insert (intercalate) between base pairs in DNA by dint of its planar configuration, Brenner and Crick ingeniously exploited this observation to decipher fundamental aspects of the genetic code (a story told in Chapter 13). Meanwhile, the so-called coding problem remained precisely that—a problem. As the decade of the fifties passed, "stuck" was a most apt descriptor of research on the genetic code. "The whole business of the code was a complete mess . . . [W]e were completely lost, you see. Didn't know where to turn. Nothing fitted," Crick lamented.[19]

Despite those frustrating times, a decade later Brenner and Benzer, together with Charles Yanofsky, received the prestigious Lasker Award for Basic Medical Research "for their brilliant contribution to molecular genetics—the science of the chemistry of heredity."[20] Brenner's wish that he and Benzer might share the stage of the Stockholm Concert Hall one December 10th never materialized. But they came close. "I would venture a guess that Sydney wanted to share a Nobel Prize with Benzer," Jim Watson confided. "He should have. I know that they were nominated together a few times."[20] Indeed, the only time that Watson and Crick conominated colleagues for the prize, their selections were Brenner, Benzer, and Yanofsky.[21]

Sydney Brenner receiving the Nobel Prize award from the King of Sweden, 2002. (Courtesy of Martin Chalfie.)

The Brenner family, circa 1952. (Left to right) Maurice Finn, husband of Sydney's sister Phyllis, Sydney's father Morris Brenner, Sydney, his mother Leah Brenner, Phyllis Finn, and Sydney's younger brother Isaac (Joe). (Courtesy of Phyllis Finn.)

The Governing Board of the MRC Laboratory of Molecular Biology, 1967. (Left to right) Hugh Huxley, John Kendrew, Max Perutz, Francis Crick, Fred Sanger, and Sydney Brenner. (Courtesy of MRC Laboratory of Molecular Biology.)

Field Marshall the Right Honorable Jan Smuts with the NUSAS Executive Committee, July 1949. (Left to right) Brenner, Patricia Arnett, Smuts, Donald Prosser, Ray Krigerand, and NUSAS President Phillip Tobias. (Courtesy of Sydney Brenner.)

Phillip Tobias and Brenner, circa 1950. (Courtesy of Sydney Brenner.)

Brenner in the South African bush, circa 1945. (Courtesy of Sydney Brenner.)

Brenner (right foreground) with the late Seymour Papert, 1972. Francis Crick is seated in the background to the left of Papert. (Courtesy of Sydney Brenner.)

In the South African Army Reserve, circa 1944. Brenner is on the extreme right, second row from the back. (Courtesy of Sydney Brenner.)

May and Sydney Brenner around the time of their wedding in 1952. (Courtesy of Sydney Brenner.)

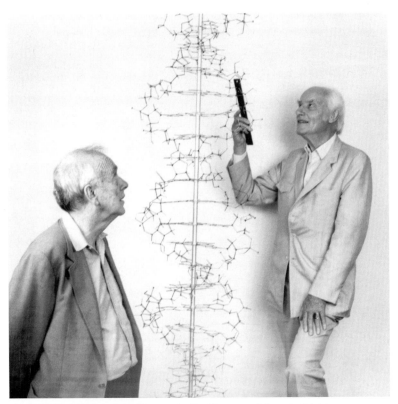

The DNA Model being viewed in 1990 by Jim Watson and Francis Crick in a recreation of the iconic photograph of 1953. (Photograph by Sue Lauter, courtesy of the Cold Spring Harbor Laboratory Archives.)

Brenner (right front) at the Cold Spring Harbor Laboratory Phage Meeting, 1954. Immediately behind him are (left to right) Al Hershey, Milislav Demerec, and Francis Crick. (Photograph by Norton Zinder, courtesy of the Cold Spring Harbor Laboratory Archives.)

Brenner and Seymour Benzer at the 1975 Cold Spring Harbor Symposium. (Courtesy of Sydney Brenner.)

Brenner (second from left) at the Cavendish Laboratory, 1957. Max Perutz is on the far right. (Courtesy of Sydney Brenner.)

The Brenner family, circa 1962. Left to right: *Belinda, Stefan, May, Jonathan Balkind, and Carla. (Courtesy of Sydney Brenner.)*

Aaron Klug and Brenner, circa 2000, Cambridge. (Courtesy of Sydney Brenner.)

Sydney Brenner speaking with Raymond Baxter (back to camera) of the BBC Television Service, late 1959. Molecular models, left to right: tobacco mosaic virus, turnip yellow virus, DNA, and model of a phage (held by Brenner). (Courtesy of Sydney Brenner.)

UNIVERSITY OF CAMBRIDGE   **DEPARTMENT OF PHYSICS**

TELEPHONE
CAMBRIDGE 54481

CAVENDISH LABORATORY
FREE SCHOOL LANE
CAMBRIDGE

M.R.C. Unit.
21st February, 1961.

Dear Jim,

    I am deeply sorry that you were upset over the last part of our paper, but we did discuss the matter with Francois Gros who did not raise any objections. I see now that it has not been phrased in the best manner and I am pleased that the out come of all this is that we shall publish simultaneously.

    I have written to the Editor of Nature saying that we wish to hold up publication so that simultaneous publication can be achieved; for this reason I would be grateful if you could send me a copy of your paper as you submit it. If you like, Max could submit your paper as he did ours.

Letter from Brenner to Jim Watson, 21 February 1961, concerning publication of their respective manuscripts on mRNA in Nature *(see Chapter 12). (Courtesy of the Cold Spring Harbor Laboratory Archives.)*

Francis Crick and Brenner at the blackboard, 1986.
(Courtesy of MRC Laboratory of Molecular Biology.)

Brenner with his daughters Belinda (left) and Carla (right) on the beach
at La Tranche, 1962. (Courtesy of Sydney Brenner.)

François Jacob and his wife Lise on the beach at La Tranche, 1962.
(Courtesy of Sydney Brenner.)

Nicol Thomson at the electron microscope, circa 1975. (Courtesy of MRC Laboratory of Molecular Biology.)

Nathaniel Mayer Victor Rothschild, 3rd Baron Rothschild by Bassano. (© National Portrait Gallery, London.)

John Sulston, circa 1985. (Courtesy of MRC Laboratory of Molecular Biology.)

Bob Horvitz, circa 1976 (left) and about 30 years later (right). (Courtesy of MRC Laboratory of Molecular Biology and http://bcrc.bio.umass.edu/hhmi/node/63.)

Paul Berg at the 1963 Symposium on Synthesis and Structure of Macromolecules. (Courtesy of the Cold Spring Harbor Laboratory Archives.)

Asilomar, 1975. (Left to right) Maxine Singer, Norton Zinder, Brenner, and Paul Berg. (Courtesy of MIT Institute Archives & Special Collections.)

Jim Watson and Brenner at Asilomar, 1975. (Courtesy MIT Institute Archives & Special Collections.)

*Jonathan Karn, circa 2000. (Courtesy of
MRC Laboratory of Molecular Biology.)*

*Brenner with Robert Horvitz and John Sulston receiving the 2002 Nobel Prize in Medicine.
(From J. Bras. Patol. Med. Lab. vol.40, No.4, Rio de Janeiro, August 2004).*

*Brenner receives a
symbolic key to the LMB
from departing director
Max Perutz. The key was
inscribed with the letters
AUG, the initiator codon
for protein synthesis.
(Courtesy of MRC
Laboratory of Molecular
Biology.)*

Celebrating Fred Sanger's second Nobel Prize, 1980. Left to right: Sanger's daughter Sally, David Secher (rear), Sanger, Brenner (with two crutches), and Max Perutz. (Courtesy of MRC Laboratory of Molecular Biology.)

Brenner talking to one of his grandchildren at a garden party. A caption on the back of the photo reads: "Would you like some champagne?" (Courtesy of Sydney Brenner.)

Some of the fugu principals. Left to right: Sam Eletr, Sam Aparicio, Greg Elgar, Brenner, Byrappa Venkatesh, Theresa Patricio Gouveia, and Alexandre Quintanilha in Oporto, Portugal, circa 2003. (Courtesy of Sam Aparicio.)

Loose ends

from Current Biology

by Sydney Brenner

Loose Ends, a collection of essays by Brenner.
(Courtesy of Vitek Tracz.)

The definition of data-mining:
what's my data is mine and what's yours is also mine.

© 2002 *The Scientist*

In the words of Sydney Brenner: Data Mining. Image by the artist Andrzej Krauze. (Courtesy of the author.)

Brenner lecturing, circa 1960. (Courtesy of Sydney Brenner.)

François Jacob, Max Bernstiel, and Brenner at the 1985 Cold Spring Harbor Symposium. (Courtesy of the Cold Spring Harbor Laboratory Archives.)

At Oxford, circa 1953.
(Courtesy of Sydney Brenner.)

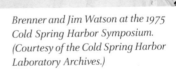

Brenner and Jim Watson at the 1975
Cold Spring Harbor Symposium.
(Courtesy of the Cold Spring Harbor
Laboratory Archives.)

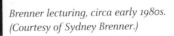

Brenner lecturing, circa early 1980s.
(Courtesy of Sydney Brenner.)

Brenner at his home in Ely, circa 2005. (Courtesy of Sydney Brenner.)

Brenner speaking at the EMBO, circa 1972. (Courtesy of Sydney Brenner.)

Leaving his footprint at the Cradle of Mankind Museum and Archive, 27 March 2008. (Courtesy of the City of Mogale, South Africa www.mogalecity.gov.za.)

10

The MRC Laboratory of Molecular Biology

I'm not the sort of person who likes to think in isolation

IN THE EARLY 1960S, URGENT EFFORTS WERE MADE TO ADDRESS the restrictive space limitations at the Cavendish Laboratories. Perutz and Kendrew's structural work on myoglobin and haemoglobin was expanding, and it became increasingly obvious that this unit, which had called itself the Unit for Research on the Molecular Structure of Biological Systems since 1947, could not continue to survive in the Cavendish. Perutz loudly protested that by 1956 the unit had grown to a size where he spent much of his time scrounging bench space in a butterfly museum here, or the abandoned cyclotron room there.[1] Some relief of the space crisis came through the generosity of David Keilin, professor of biology and parasitology at Cambridge, whose interest in Perutz's work prompted him to donate laboratory facilities at his Molteno Institute. In fact, from 1938 until the early 1950s, protein chemistry was conducted at the Molteno, while X-ray work endured at the Cavendish, with Perutz busily scurrying back and forth between the two.[2] In 1956 the university provided use of the rapidly decaying Metallurgy Hut (affectionately referred to by the staff simply as "the hut") located next to the Cavendish unit.

Considerable thought was also given to relocating Brenner, Crick, and their respective research teams elsewhere in the university, perhaps in the Departments of Genetics or Chemistry. Indeed, in the late 1950s, Crick formally applied for the Arthur Balfour Chair of Genetics recently vacated by the famous statistician and population geneticist R. A. Fisher, with the explicit intention of moving Brenner with him. Fisher, who was well acquainted with Crick, supported his candidacy. But the search committee courted the Italian geneticist Guido Pontecorvo. After withdrawing his candidacy (because he did not feel especially welcomed in Cambridge), Pontecorvo informed Crick that he had told Sir James Gray, professor of botany at Cambridge, that "if they have any sense left they should appoint you."[3] However, the Oxford cytogeneticist Cyril Darlington persuaded the search committee that "Crick could not claim to be a geneticist"[4]—a pronouncement that revealed far more about the level of interest in molecular genetics than about Francis Crick. In retrospect, had Crick been offered the position, taking Brenner with him, the many rewards earned by the mix of scientists at the Laboratory of Molecular Biology might never have been realized.

As the space situation worsened, Keilin suggested to Bragg that he approach the Secretary of the Medical Research Council (MRC), Sir Edward Mellanby, about the problem. In traditional fashion Bragg lunched with Mellanby at the Athenaeum Club, an exclusive London gentleman's club situated next to the Royal Society on a quiet street off the Mall. Mellanby was responsible for deploying the MRC's scientific resources during World War II and for reconstructing and notably expanding the organization thereafter. He proved to be a sympathetic listener. Bragg stressed the enormous potential of the work being initiated by Crick and Brenner at the Cavendish. Encouraging words from his friend David Keilin may also have influenced Mellanby. In the end the MRC Secretary promised a new building forthwith.

When informed about the welcome news, Brenner and Crick scoured the Cambridge University campus for a suitable location. But no appropriate site was available. Nor was the situation helped by the ensuing political intrigue, mainly centered on opposition by many on the Cambridge University faculty to the establishment of a freestanding research institute that was financially independent. Faced with the reality that they could not survive unless they expanded physically, Crick, Brenner, and their Cavendish colleagues switched their focus to a new building site on Hills Road, several miles from the university campus, but just across the street from the future home of the Addenbrooke's Hospital, which was to become a teaching hospital of Cambridge Medical School. No sooner had the new laboratory opened than a proposal was drafted for yet further space, and in 1968 a 13-bay extension of the new building was added. Other bits of space continued to accrue over the years, including a new lecture theater named for Perutz in 2002.[5]

The physical facility was no architectural masterpiece. A journalist for the *Sunday Times* wrote: "It would be difficult to find a less distinguished building. It lies five stories high in anonymous sixties brick on the outskirts of Cambridge, looking rather like the branch office of an insurance company."[5]

In the spring of 1962, Queen Elizabeth II ceremoniously opened the new MRC Laboratory of Molecular Biology on Hills Road. The Queen expressed appropriate interest in Perutz's tour of the facility, charming even the anti-royalists dutifully in attendance. Her ladies were, of course, also in tow, and when shown a stick and ball model of a protein, perhaps hemoglobin, one of them is reported to have exclaimed, "I had no idea that we had all these little coloured balls inside us."[6]

The antiroyalist contingent present for the occasion did not include Brenner and Crick, who symbolically absented themselves from the occasion. However, years later during Brenner's watch as director of the LMB, he endured a visit from Prime Minister Margaret Thatcher. This visit was preceded by a phone call from a gentleman who announced himself as Mr. Nick Sebastian, from Downing Street. As well as the famous street of that name in London, there is a Downing Street in Cambridge, which is reasonably well known for a clutch of laboratories situated there. Brenner was an occasional recipient of phone calls from some of these laboratories, requesting help of one sort or another. So he politely asked Mr. Sebastian which lab he represented. When the gentleman firmly repeated that he was located on Downing Street, the now irritated Brenner stated, "Yes, I heard what you said. But which laboratory are you from?" After a prolonged pause Mr. Sebastian very clearly stated, "Number 10 Downing Street," at which point the usually unflappable Brenner exclaimed apologetically, "Oh my God. Is that the one you mean?" The gentleman was, in fact, calling to inform Brenner that Mrs. Thatcher wished to visit the LMB and wanted some preparatory reading matter.

The change in name from the Laboratory for the Study of Molecular Structure of Biological Systems to the Laboratory for Molecular Biology was both enterprising and calculated. Warren Weaver, Director for Natural Sciences at the U.S.-based Rockefeller Foundation, is credited with coining the phrase "molecular biology" some 30 years earlier. In the early 1960s the term (still new to many) was an apt descriptor of Brenner and Crick's aspirations and intentions for the new research unit. Among the 25 research scientists then at the LMB, one (Fred Sanger) was a Nobel Laureate and four were elected Fellows of the Royal Society. Within months of the laboratory's opening, three more Nobel Laureates and two more Fellows emerged, confronting the new Secretary of the MRC, Sir Harold Himsworth, with a challenging group of scientific prima donnas.

Himsworth suggested to LMB director Max Perutz that he give each of the senior scientists responsibility for the policy of his/her own section or division. Perutz heeded this advice and drafted a constitution for the new laboratory that established three autonomous divisions: Protein Crystallography (later called Structural Studies) headed by John Kendrew, Protein Chemistry headed by Fred Sanger, and Molecular Genetics headed by Crick.[7] These individuals, together with Perutz, constituted a Governing Board that allocated financial resources to the various divisions. (Perutz adamantly refused the formal title of Director, preferring instead to be designated as Chairman of the Governing Board.) Subsequently, a subgroup of the Structural Studies division was established, headed by Aaron Klug, who joined the LMB in 1962, and Hugh Huxley, a more recent recruit. Klug, Huxley, and Brenner (who was appointed cohead of Crick's Division) were persuaded by Perutz to join the Governing Board.[8]

Aaron Klug and Sydney Brenner share extraordinary parallels in their scientific careers. Born in South Africa just a year before Brenner, Klug is also the son of Jewish immigrant parents from Eastern Europe, and he, too, attended Wits medical school, but dropped out after his second year to pursue a career in physics at the University of Cape Town. Like Brenner, Klug left South Africa with an 1851 Exhibition Scholarship and obtained his Ph.D. at Cambridge University, before joining J. D. Bernal's X-ray crystallography group at Birkbeck College. There he met Rosalind Franklin,[a] whose spark kindled his interest in viruses.

Remarkably, Brenner and Klug both pursued the bulk of their scientific careers at the LMB, and both men, now in the eighth decade of their lives, carry the distinction of Nobel Laureate. Without his fellow South African's flamboyance, Klug would in due course succeed Brenner as director of the LMB, a succession, as Chapter 21 will reveal, that was not without some acrimony.

Each division of the LMB had its own floor with a common workshop and store for supplies. Division heads could, in theory, appeal directly to Himsworth if disputes arose over the sharing of resources. But as a matter of courtesy, such matters were always passed through Perutz. This administrative structure was a significant departure from usual MRC practice, which typically provided long-term funding to a single director to build a group that

---

[a] Before joining Bernal's group, Rosalind Franklin was a structural biologist at King's College who worked on the X-ray structure of DNA. Though well thought of as an experimentalist, she worked very independently and did not exactly see eye-to-eye with her King's College colleague Maurice Wilkins, nor, for that matter, with Watson and Crick. Watson portrayed Franklin rather negatively in his famous book *The Double Helix*, but later revised his views and readily concedes that Franklin would almost certainly have deciphered the DNA structure herself had she not been upstaged by the Cambridge pair. Franklin died of ovarian cancer as a young woman.

he or she would oversee. Of course, this unwritten edict was precisely what Perutz wanted to finesse.

As plans for the nearby Addenbrooke's hospital matured, the MRC announced that a single cafeteria could easily cater to the needs of both the hospital and LMB personnel. Perutz would have none of that. Convinced of the imperative of a dedicated facility in the LMB where people could congregate and exchange ideas while enjoying a brief respite, he established a canteen or tearoom. The tearoom soon became a focal gathering place for the scientific staff and their trainees. Visiting American scholars were surprised at the regularity with which people frequented the room for a cup of tea, at almost any time of day or night. They quickly became eager converts, both to the routine and to English tea.

To further promote scientific interaction, Francis Crick instituted a week-long seminar format (often referred to as "Crick week") at the beginning of each academic year, during which investigators presented their research. The entire research staff, including post-doctoral fellows and graduate students, was required to attend these presentations. The quality and quantity of these in-house seminars reflected the astonishing productivity of the laboratory during its hey-day in the 1960s and 1970s. "The symposia and seminars presented at the LMB during Crick week were extraordinary," former postdoctoral fellow Jonathan Karn (who joined the laboratory in 1977) stated. "I well remember talks from Fred Sanger on his dideoxy method for DNA sequencing, from César Milstein on the first monoclonal antibodies, Nigel Unwin describing electron microscopic techniques for determining three dimensional structures, Aaron Klug reporting on higher order chromatin structures—and on and on."[9] To be sure, anyone lucky enough to listen in on the whirlwind of talks presented during Crick week at the LMB was likely the beneficiary of news about some of the most cutting-edge research in the world.

Speakers were admonished by Crick to pitch their talks at a level that the entire scientifically diverse audience could comprehend.[10] Purportedly, during one such seminar, Perutz leaned over to Brenner and in a quiet whisper asked for clarification of a statement by the speaker. Brenner immediately leapt to his feet. "Stop the seminar. Stop the seminar," he announced. "Max doesn't understand something." Perutz presumably did not share in the ensuing amusement. While clearly no intellectual slouch, Perutz sometimes had trouble keeping up with the "quick-fire punning and repartee"[11] that characterized many conversations in the hallways and cafeteria.

However, Perutz was by no means stodgy. He ushered in a refreshing new era at Cambridge, one that reflected a distinct departure from the historically stuffy traditions of British science. The buttoned-up white lab coats and other symbols of formality, such as the use of titles, disappeared, and the general tone of the LMB was relaxed and open. His management style was essentially hands-off. He empowered his division heads with considerable independence and authority, attending to their scientific needs with remarkable efficiency and alacrity. As we shall see in Chapter 20, while his supportive attitude endeared him to many, his spending was poorly managed and Perutz eventually ran the LMB into financial debt. Nonetheless, he was well liked by the MRC leadership, which observed the rapid ascendancy of the scientific reputation of the LMB with pride and satisfaction. In October 1962, when telegrams arrived announcing that Nobel Prizes had been awarded to Perutz and Kendrew,[b] in addition to Watson and Crick, the champagne flowed. The MRC added an extension to the LMB, and another appeared towards the end of the 1970s.

The reputation of the lab soared and, for many years—certainly well into the 1980s—the LMB enjoyed a revered status as the premier molecular biology research and training institution in the world. Applications for post-doctoral positions were bountiful, and paid fellowships for these were at a premium. Many students and post-doctoral fellows were intent on working with Brenner or Crick. Those who could not be accommodated in their crowded laboratories enjoyed superb mentoring from other members of the permanent staff and benefited from frequent interaction with them in the halls and corridors, and in the tearoom, of course. An impressive number of these trainees went on to distinguished scientific careers, and a substantial fraction was elected to the Royal Society of London or the U.S. National Academy of Sciences. More than a few (Martin Chalfie, Robert Horvitz, Roger Kornberg, Sidney Altman, and Andrew Fire) became Nobel Laureates.

With few exceptions these individuals considered their stay at the LMB as one of the highlights of their scientific careers, especially the Americans, many of whom were visiting England for the first time. To a man (and woman), they hold Brenner in the highest regard. Though aware of his sometimes brash and dismissive manner, and his inclination to biting humor, their respect and affection for Brenner prompted a spontaneous tribute in the

---

[b] This was the first (and only) occasion to date that two research groups from a single laboratory were awarded the Nobel Prize in the same year. The awards to the Laboratory for Molecular Biology did not end then. Nobel Prizes were awarded to Fred Sanger in 1980 (his second), to the South African Aaron Klug in 1982, to César Milstein and George Kohler in 1984, to John Walker in 1997, and, of course, to Brenner and his Cambridge colleague John Sulston in 2002.

summer of 1983, by way of a special symposium. What began as a relatively small affair featuring some of Brenner's former post-doctoral fellows and students, progressed to a larger event honoring other senior figures at the LMB. Participants at this memorable occasion, many of whom journeyed from the United States at their own expense, wrote endearing letters of thanks to their former mentor. These were compiled as a bound volume and presented to Brenner by former post-doctoral fellows and symposium organizers, Paul Wassarman and Malcolm Gefter. Art Landy, now a professor at Brown University, expressed sentiments shared by many:

> It is pretty much agreed among the alumni that we have not found anywhere else such an intellectual-social environment. We all know (and now especially appreciate) that while the environment always felt like something that just happened, in fact it took a lot of insight to create and considerable effort to maintain. The short hallways, open doors and daily tea owed their significance to the people we could talk to.

> I must confess that my interaction with you had a certain element of tension. I felt like a moth attracted to the brilliance of a light bulb despite the threat of getting burned. It was exciting. Your dedication to challenging assumptions on every topic (from traffic control in Cambridge to fiscal policy in Southeast Asia) dominated every conversation and bull session. Even more unique, and certainly more significant, was the string of original or provoking (and sometimes outrageous) alternatives you always came up with.[12]

Malcolm Gefter, now a professor emeritus at MIT, wrote:

> I remember the days at the MRC more for the long lunch hours, coffee breaks and Saturday mornings than I do for the time spent at the lab bench. . . . One of the greatest lessons I learned was not how to do an experiment but how not to do one. . . . [Experiments] were the embodiment of an intellectual exercise and not a vehicle for filling a notebook or for a publication, nor were they the vehicle to achieve fame or to beat the competition. Thus, before doing an experiment at the MRC, it had to pass the rigorous intellectual challenges conducted over coffee. What survived these challenges were only the best experiments you could think of, amended by the jurors (your colleagues). If you survived coffee, how could you not do an outstanding piece of research?[12]

Life in the research laboratory was rarely dull for Brenner, regardless of which laboratory or which era one considers. However, his stay in Cambridge, starting in the early 1960s, was one that he speaks of with especially fond memories. "The new lab in Cambridge was absolutely the most marvelous place to work in. Francis and I had diverged along different paths. But we continued to share an office that we equipped with blackboards. We still enjoyed using chalk then! We had three blackboards and we used them on a daily basis."

Brenner adopted an accommodating attitude with his graduate students and post-doctoral fellows. Fundamentally, his notion of supervision was to leave well alone—and wait to see what emerged—not that he was unavailable or unapproachable about research problems and progress. But he was more likely to be fruitfully engaged in the corridors or the tearoom than by formal appointment in his office. With those in his laboratory, he espoused a sink-or-swim philosophy; if one did not soon mature into an independent and productive investigator, one had better seek pastures of a different color, green or otherwise. At one point Brenner orchestrated a forum dubbed the SB Educational Society, in which post-doctoral fellows were required to discuss their work at five minutes notice. This was not much liked by the American fellows who hated being caught off guard and wanted time to prepare their discussions. "But it always seemed to me that if you had to prepare anything in order to discuss what you were doing at the moment you'd better get worried about that, because it meant that you didn't know what you were doing."

A less formal discussion group meeting was simply referred to as "Saturday morning coffee," during which people congregated in the kitchen on Saturday mornings around coffee (or tea) and discussed all sorts of topics, both scientific and nonscientific. Coffee-time in the morning or tea-time in the afternoon was always an opportunity to talk science with anyone who happened to be passing through. In particular, after lunch, Brenner would hold court, talking for an hour or more with the young trainees.[13] He has stated:

> I believe that this type of on-going conversation is critical in science. I'm not the sort of person who likes to think in isolation. When an idea germinates in my mind I know that it's at least fifty percent wrong initially. And it's only in playing with the idea by bouncing it off others that one can refine it and retain what's essential. I believe that sort of interaction, during which ideas become progressively clarified, is one of the great joys of science. The other of course is to get into the lab and translate the idea into real experiments— and get the right answers.

11

Messenger RNA—The Concept

The paradox of the prodigious rate of protein synthesis

T HE DISCOVERY OF MESSENGER RNA (mRNA), the nucleic acid template for
the assembly of amino acids into polypeptide chains, was one of several
spectacular advances in molecular biology in the early 1960s. Brenner was
seminally involved in this momentous episode, one that clarified a central
issue in protein synthesis—how coding information in DNA is translated to
amino acids in proteins.

Brenner's association with mRNA is rooted in observations that emerged
from his quest to identify phage gene-protein pairs that might yield the
much-anticipated colinearity between the coding region of genes in DNA
and amino acids in the proteins they encode. As we saw in Chapter 9, during
the furious (and ultimately futile) search for tractable phage gene-protein
pairs, Brenner alighted on several bacteriophage-encoded proteins, notably
tail fibers, sheath proteins, and head proteins. He was particularly struck
by the enormous quantity of head proteins expressed when phages infect their
bacterial hosts. Brenner euphemistically dubbed this observation "the paradox
of the prodigious rate of protein synthesis." As the decade of the 1960s
unfolded, the dogma was that if RNA indeed spelled out a code for the assembly of amino acids into polypeptides, it was the RNA in ribosomes that fulfilled
this function. Since there were no indications that new ribosomes are synthesized during phage infection, Brenner found the huge burst of phage-induced
protein synthesis intriguing. "One had to seriously consider the possibility that
a small amount of new ribosomal synthesis did in fact transpire during phage
infection. But if so, it directed a vast amount of protein synthesis."

RNA synthesis in phage-infected cells had been cursorily examined in
several laboratories. We saw in Chapter 7 that Al Hershey is remembered
for the classic Waring blender experiments he conducted with his young

colleague Martha Chase. However, in 1953 he published a little known paper that identified what later became known as messenger RNA. In brief, while investigating DNA synthesis in phage infected *Escherichia coli*, Hershey used radio-labeled phosphate to monitor the synthesis of nucleic acids and parenthetically noted the rapid incorporation and turnover of radioactivity associated with RNA. He commented on this observation but offered no speculation as to its significance: DNA, not RNA, was the focus of Hershey's interest then.

> [I]t was noted that an extremely rapid incorporation of P^{32} into RNA occurred. . . . Enzymatic tests showed that the active material was authentic RNA. . . . Its metabolic activity clearly distinguishes this material from the bulk of RNA that was formed in the cell before infection.[1]

Several years later the transient synthesis of RNA following phage infection of bacterial cells resurfaced in a more provocative context from an unexpected quarter. The intense American spotlight on building an atomic bomb during World War II prompted the establishment of several national laboratories under the aegis of the newly constituted U.S. Atomic Energy Commission. Above and beyond the primary charge of developing a nuclear arsenal, these laboratories—established at Oak Ridge, Tennessee; Brookhaven, New York; Los Alamos, New Mexico; and on the outskirts of Chicago at the Argonne National Laboratory—were charged with the study of radiation effects on humans with the primary objective of mitigating the harmful consequences of radiation exposure on the battlefield. This mission launched the discipline of radiobiology, a field that ultimately matured to embrace the collective biological responses of living cells to DNA damage.

Elliot Volkin, a biochemist in the Biochemistry Section in the Biology Division of the Oak Ridge National Laboratory, was examining the effects of radiation on nucleic acids. Mindful of the sophisticated chromatographic methods for resolving radioactive products of nuclear fission developed by Waldo Cohen and others at the Oak Ridge National Laboratory, Volkin exploited this technology to resolve the nucleotides of RNA and DNA.[2]

Like Hershey, Volkin added T2 bacteriophage to a bacterial culture that was spiked a few minutes later with radioactive inorganic phosphate to label newly synthesized nucleic acids. After isolating RNA and DNA he converted the RNA to mononucleotides and analyzed the composition of the radio-labeled fraction. Volkin documented the reproducible uptake of small amounts of ^{32}P into RNA after phage infection. "What gave me an immediate jolt," he wrote in a 1995 retrospective of these experiments, "was the observation that the total [radioactivity] counts and the specific activities associated with the individual nucleotides did not resemble in any way the nucleotide composition of the host *E. coli* DNA. . . ." Volkin co-opted his Oak Ridge colleague Lazarus "Larry" Astrachan to investigate this observation more deeply. Further experiments realized the conclusion and Volkin noted:

> . . . that a previously unrecognized kind of RNA is synthesized *de novo* immediately upon lytic phage infection. The RNA . . . has a base composition virtually identical to that of the infecting phage's DNA. It undergoes rapid metabolic turnover, is associated with a qualitatively minor subcellular component, and its activity appears to be interconnected with protein synthesis.[3]

Volkin and Astrachan subsequently reported essentially identical results following infection of *E. coli* with phage T7,[4] indicating that this burst of RNA synthesis was not phage T2-specific.

The summer of 1956 was marked by a series of meetings on the emerging field of molecular biology. Following the annual Cold Spring Harbor Symposium, a meeting on the chemical basis of heredity entitled Genetic Mechanisms: Structure and Function convened at the McCollum-Pratt Institute at Johns Hopkins University in Baltimore. The annual Gordon Conference on Nucleic Acids and a conference at the University of Michigan in Ann Arbor in turn followed these. Volkin presented his and Astrachan's findings at the Baltimore meeting. But neither they nor for some years anyone else realized that the RNA they had discovered was, in fact, phage DNA-directed messenger RNA. Neither Crick nor Brenner heard Volkin's presentation. Crick joined the meeting circuit later—and Brenner was still in South Africa.

Much has been made of this missed opportunity by Astrachan and Volkin. Commenting on the fact that Crick was not at the Baltimore meeting and apparently first heard about DNA-like RNA in phage-infected cells when he read the conference proceedings published in 1957, historian Horace Judson disparagingly noted: "[The Volkin and Astrachan RNA] could not quite be dismissed as the inconsequential and self-admittedly sloppy work of people not well known at a lab not highly regarded."[5] But, the notion of "self-admittedly sloppy work" is not strictly accurate. According to Judson, Volkin emphasized that "the experiments had met technical difficulties [and] the results might not yet be clean."[5] However, Volkin and Astrachan's publication in the proceedings of the meeting (entitled "RNA Metabolism in T2-Infected *Escherichia coli*") merely makes two cautionary references to possible technical difficulties, neither of which begged obvious concern in their minds about the validity of their data.

Nor is it apparent that Volkin and Astrachan were summarily dismissed as "people not well known." On the contrary, in an interview at his home in 2003 Volkin stated: "I can well remember sitting on the lawn at Cold Spring Harbor and telling Sydney Brenner (who, in 1956 was in the midst of his extended stay there as a Carnegie fellow) about our experiments. I gave a presentation on our RNA research to the phage group there."[6] Volkin also recalled that "Salvatore Luria convinced him and Astrachan to publish his results in the *Journal of Virology*."[6] When reminiscing about the 1956 annual meeting of the Federation of Societies for Experimental Biology, the Berkeley geneticist Thomas H. Jukes later recalled that he had "squeezed his way into a doorway of a packed room to hear a paper by Volkin and Astrachan on DNA-like RNA."[7] Volkin also presented the work at an international microbiology meeting in Stockholm and at the International Biochemical Society meeting in Vienna in 1958.[4] Therefore, although neither the Biology Division at Oak Ridge in general, nor Elliot Volkin or Larry Astrachan in particular, enjoyed international reputations in the emerging discipline of molecular biology, the notion that their intriguing but confounding observations were summarily dismissed is an exaggeration.

More cogent explanations for this intellectual lapse come readily to mind. For example, the observations by the Oak Ridge team were published over a span of three years, and none of the papers explicitly stated the salient observations summarized by Volkin with the benefit of 20/20 hindsight. Their frequent allusion to the possibility that the new RNA observed after phage infection may be required for the synthesis of phage DNA was another confounding issue. Indeed, in one of these papers, the Oak Ridge pair disconsolately concluded: "Questions about how and why the pattern of isotope incorporation changes after phage infection, and why the pattern after infection is similar to the base composition of bacteriophage DNA are unanswered."[8]

Al Hershey and Volkin and Astrachan were not the only contemporary scientists to miss the messenger RNA boat. In 1960 the Lithuanian-American microbiologist Martynas Ycas and his colleague W. S. Vincent reported [in no less a journal than the widely read *Proceedings of the National Academy of Sciences* (PNAS)] the identification of an RNA fraction from yeast with a base composition similar to that of yeast DNA.

The existence of a species of RNA related in composition to DNA has been previously reported by Astrachan and Volkin in the special case of *E. coli* infected with bacteriophage. Like our active fraction from yeast, it shows a rapid rate of turnover. . . . The function of such an RNA fraction is not yet clear. In view of its composition *it might be a primary gene product, acting as an agent for transmission of genetic information from DNA to protein* [author's italics].[9]

The scientific world was exasperatingly close to discovering messenger RNA in the period between the mid-1950 and early 1960s. But the intellectual leap required to connect RNA—known to be associated with ribosomes and widely believed to be the template for protein synthesis—to isolated reports of a tiny fraction of RNA rapidly turning over in bacteriophage-infected *E. coli* was simply too great. François Jacob and Jacques Monod, of the famous school of phage geneticists led by Andre Lwoff and Elie Wollman at the Pasteur Institute in France, came tantalizingly close to inferring the existence of mRNA while carrying out their brilliant studies on the regulation of expression of the gene for β-galactosidase in *E. coli*. (They shared the Nobel Prize for physiology or medicine in 1965 for these studies.) However, they, too, maddeningly failed to make the connection. The truth is that no one explored the implications buried in the studies by Hershey, Volkin and Astrachan, Ycas, and Jacob and Monod—until a fateful day in 1960, when a small group gathered to brainstorm in King's College, Cambridge.

In the spring of 1960, the Society for General Microbiology held its 10th symposium in London. Following the meeting, François Jacob, Ole Maaløe (a Danish member of the international phage group), and Alan Garen (from the Massachusetts Institute of Technology) took a train to Cambridge to spend the long Easter weekend with Brenner and Crick. On Good Friday, April 15, the visitors met in Brenner's rooms in King's College with Crick, Brenner, and one or two others whose names no one seems to remember.[10] Brenner recalled that the drama unfolded "when we started discussing Jacob's and Monod's PaJaMo experiments, [so-called because they were executed by the trio of the American Arthur (Art) Pardee (Pa), and the Frenchmen François Jacob (Ja), and Jacques Monod (Mo)]."

Akin to the intellectual partnership between Crick and Brenner, François Jacob and Jacques Monod of the Pasteur Institute in Paris joined forces in the late 1940s to pursue their own curiosities about the intricacies of gene function in bacteria.[11] Jacob related his celebrated collaboration with Jacques Monod in his autobiography *The Statue Within*, published in 1988.

In the fall of 1957 . . . Jacques and I decided . . . to work together more closely on . . . the metabolism of lactose, milk sugar, in the colon bacillus [*E. coli*]. The first series of experiments suggested itself. . . . To use lactose, the colon bacillus employs the very stable enzyme galactosidase. . . . To check out the system, to regulate it, to control it, we decided to . . . have a "high frequency" male inject the gene of galactosidase into a female possessing a mutation that prevented it from producing this enzyme; then to find out whether the injected gene functioned in the female; whether the enzyme was synthesized; after how long a time; under what conditions. . . . First experiment, first surprise. A big one! The gene functioned immediately, with no delay. As soon as it entered, a synthesis of the enzyme was observed at the maximum rate, at full speed. A puzzling result which did not fit with generally received ideas for explaining the synthesis of proteins.[12]

Jacob and Monod were singularly (and frustratingly) focused on comprehending how expression of the β-galactosidase gene was regulated. As every college biology major now knows, the solution lay in elucidating the regulation of gene repression and induction, an explanation that eluded them for longer than they wished, but that ultimately won them a Nobel Prize. At the heart of this regulatory system, and central to the present discussion, was a vital clue to the existence of messenger RNA.

We knew that protein synthesis took place in the cytoplasm, on tiny granules called "ribosomes." . . . But these ribosomes, made up of proteins and RNA, were very stable structures, lasting for several generations, a scheme that [did not] accord . . . with the synthesis of galactosidase immediately upon entry of the gene. . . . [We wavered] between . . . two possible hypotheses: either direct synthesis of the protein on DNA itself, with no intermediary; *or production of an unstable intermediary, probably an RNA with rapid renewal* [author's italics]. But the former hypothesis seemed highly improbable and the latter without a chemical basis, without any trace of a molecule that could substantiate it.

I had related this story, and our doubts, at a small colloquium on microbial genetics organized in Copenhagen by Ole Maaløe in September 1959. A small group attended, including notably Jim Watson, Francis Crick, Seymour Benzer, Sydney Brenner, Jacques [Monod], and even the physicist Niels Bohr. Courteous as ever, Jim Watson spent most of the sessions ostentatiously reading a newspaper. So, when it came time for him to speak, everyone took from his pocket a newspaper and began to read it. [Brenner had distributed these just prior to the session!] I had been assigned to lay forth the views of the Pasteur group. . . . I stressed the difficulty of reconciling these events with the current model of protein synthesis. I invoked the other two hypotheses by stressing the need for an unstable intermediary, which I called X. No one asked a question. Jim continued to read his newspaper.[13]

In retrospect, it is unfortunate that Jacob raised the possibility that induced protein synthesis might directly utilize DNA as a template. Conceivably, this improbable suggestion remained foremost in the minds of at least some, if not all, of his Copenhagen audience. "To me this was a horrid possibility," Watson wrote years later,[14] suggesting a conscious or subconscious rejection of Jacob's alternative hypothesis. Brenner commented: "They seemed to have firmly excluded the possibility that new ribosomes were made to carry out this rapid protein synthesis, or if they were, they were a very small fraction of ribosomes capable of extensive synthesis, the very conundrum that we grappled with in phage-infected bacteria."

Brenner has no recollection of Jacob's mention of an unstable intermediary dubbed X at the September 1959 colloquium in Denmark. Perhaps he was too busy savoring the delight of his newspaper prank on Watson. Regardless, the electrifying epiphany that was shortly to emerge in his Cambridge rooms did not register months earlier in Copenhagen. But there is no conjecture about what transpired on that Good Friday in 1960. François Jacob picks up the story.

> Francis and Sydney wanted to discuss in detail our experiments. . . . I had new results to report: an experiment long prepared in Paris and recently completed in Berkeley by Arthur Pardee and his student Monica Riley. They had succeeded in charging the DNA of male bacteria with radioactive phosphorus; in making them transfer to females the gene of galactosidase; in letting them synthesize the protein for some minutes and then in destroying the gene through the disintegration of the radioactive phosphorus. The result was clear: once the gene was destroyed, all [protein] synthesis stopped. No gene, no enzyme. Which excluded any possibility of a stable intermediary.

> At this precise point, Francis and Sydney leaped to their feet. Began to gesticulate. To argue at top speed in great agitation. A red-faced Francis. A Sydney with bristling eyebrows. The two talked at once, all but shouting. Each trying to anticipate the other. To explain to the other what had suddenly come to mind.[15]

Brenner volunteers that the Paris group had not reflected on the relevance of the Volkin–Astrachan RNA because they were not studying phage biology. He and Crick were, but the connection was not made until all was laid out on that fateful Good Friday. "Suddenly it occurred to me that the

Volkin–Astrachan RNA must be what we called at that meeting 'messenger RNA'.[a] I got very excited and began shouting, 'Volkin–Astrachan; information intermediate; it's short-lived; a short-lived intermediate. It must be. Look at the way it turns over in phage!' Nobody knew what I was talking about because of this sudden jump in logic. But at that moment it suddenly became clear to me." Years later Brenner told Horace Judson, "suddenly I was talking to Francis, and no one else was following me. He picked it up straight away. I mean, that's the way it suddenly hit me between the eyes, that of all things, ... there had to be an RNA which was *added* to the ribosomes."[16] Crick's recollection of this critical moment was that Brenner let out a loud yelp. He had seen the answer. "The sudden flash of enlightenment when the idea was first glimpsed ... was so memorable that I can recall just where Sydney, François and I were sitting in the room when it happened."[17]

In later years Jacob sharply rebuked himself for his failure to make the connection between his and Monod's experimental observations on regulation of the β-galactosidase gene and mRNA.

> As had been shown by the American researchers, Elliot Volkin and Lazarus Astrachan, the only RNA then synthesized had two remarkable properties; on the one hand, unlike ribosomal RNA, it has the same base composition as DNA; on the other hand it renewed itself very quickly. Exactly the same properties for what we called X, the unstable intermediary we had postulated for galactosidase. Why, in Paris, when we were looking for a support material for X, had we not thought of this phage RNA? Why had I not thought of it? Ignorance? Stupidity? Oversight? Misreading of the literature? Failure of judgement?[18]

"If you talk about two things simultaneously, you have a lot of green balls bouncing and you have a lot of red balls bouncing," Brenner commented, referring metaphorically to the simultaneous discussion about the PaJaMo and the Volkin–Astrachan experiments. "Sometimes, if you're lucky, you just see just one set of balls bouncing—the correct set. For me the curtain lifted at the moment I realized that there was this unstable intermediate involved in protein synthesis." Brenner also noted the unfortunate early description of messenger RNA as being unstable. It happens to be unstable in bacteria and in the case of Jacob and Monod's experiments the notion of

---

[a] Brenner may have used the term "messenger RNA" at this informal meeting. However, the phrase was first formally introduced into the literature by Jacob and Monod in a paper entitled "Genetic Regulatory Mechanisms in the Synthesis of Proteins," *J Mol Biol* **3**: 318–356, 1961.

instability was crucial to explain how quickly the induction of β-galactosidase was shut down. "But this was not the necessary concept. The necessary concept was that you added something to a machine that gets instructed by it."

For a brief period Brenner and Crick referred to "tape RNA" because he and Crick thought of the ribosomes as a tape player that played a message when "tape RNA" was fed into it. "But then we began calling it messenger RNA because this hypothetical short-lived intermediate carried a specific message."

Thus emerged the crucial notion of an intermediary in protein synthesis equipped with coding information from genes, which carried this information to ribosomes where it was translated to the assembly of specific amino acids in polypeptide chains. But was this notion correct?

12

Messenger RNA—The Validation

It's the magnesium! It's the magnesium!

T HE EXCITEMENT OF THE DAY'S REVELATIONS CARRIED OVER into the evening. Francis Crick had arranged another soiree at the Golden Helix (his house in Cambridge) for which he was famous (notorious may be more apt). Most of the guests, many from the Cambridge intellectual and artistic community (Odile Crick was an accomplished artist) were enjoying the spread of food and wine and generally socializing. But Brenner and Jacob had other priorities. They spent much of the evening discussing the ideas that had emerged earlier in the day and planning experiments to put the messenger RNA hypothesis to the test. François Jacob remembered the occasion well.

> That evening, the Cricks were giving a party. A very British evening with the cream of Cambridge, an abundance of pretty girls, various kinds of drink, and pop music. Sydney and I, however, were much too busy and excited to take part in the festivities. We isolated ourselves in a corner with beer and sandwiches to lay our plans. . . .
>
> Since the morning session at King's College we had been forming a new representation of protein synthesis. The ribosomes had lost all specificity. They had become simple machines for assembling amino acids to form proteins of any kind, like tape recorders that can play any form of music depending on the magnetic tape inserted in them. In protein synthesis, it was X, the unstable RNA copied on a gene, that had to play the role of the magnetic tape, associating with the ribosomes to dictate to them a particular sequence of amino acids corresponding to a particular protein.[1]

Crick, too, was deeply immersed in the exciting conceptual breakthrough. If not that evening, certainly in the next few days, he dashed off a draft of a manuscript that highlighted the essential features of translation of the genetic code to proteins. The manuscript was never published, but Horace Judson read a draft, "written at a sitting, impetuous and colloquial, putting in order the arguments and evidence of the previous hours."[2] Crick's draft manuscript concluded:

> We now boldly combine the conclusions of the two experiments (Pardee, Jacob, and Monod; Volkin and Astrachan) and generalize them to produce the following hypothesis.
>
> (1) genetic RNA has the same over-all base ratios as genetic DNA,
>
> (2) it passes into the ribosomes, but it is only a minor component (10–20%?). The major part of ribosomal RNA in not genetic RNA.
>
> (3) genetic RNA is (at least in some circumstances) unstable, that is, it may only have a limited life.[2]

After much discussion Brenner and Jacob decided that the key to elevating messenger RNA from a hypothesis to a documented scientific fact lay in demonstrating that new ribosomes were *not* synthesized after phage infection. Instead, new messenger RNA was added to extant ribosomes. They agreed that this experimental goal might be achieved by exploiting the switch from bacterial protein synthesis to new phage protein synthesis following phage infection of bacteria.

How were they to demonstrate this notion experimentally? Brenner knew about the density transfer experiments that Matt Meselson and Frank Stahl had recently executed using DNA labeled with the heavy isotopes C^{13} and N^{15} to demonstrate the semi-conservative nature of DNA replication.[a] He and Jacob now decided to borrow a page from Meselson and Stahl's book. The plan was to label bacterial ribosomal proteins with heavy isotopes of carbon and nitrogen and utilize density gradient centrifugation to show that ^{32}P radio-labeled phage RNA associated entirely with these heavy ribosomes—and not with putative newly synthesized ribosomes.

Meselson, then at Caltech in Pasadena, was the only likely source of heavy carbon and nitrogen. Remarkably, independent of their new interest in mRNA, both Brenner and Jacob had been invited (by Meselson and Delbrück, respectively) to spend part of the summer of 1960 at Caltech, although specific dates and research projects had not been decided. Delighted with this fortuitous opportunity, Brenner and Jacob determined to execute these invitations at the earliest opportunity. On May 7, 1960, Brenner wrote to Meselson.

---

[a] Details of these experiments are found in Holmes FL. 2001. *Meselson, Stahl, and the replication of DNA: A history of the most beautiful experiment in biology.* Yale University Press, New Haven, CT.

Dear Matt,

This letter is to tell you about exciting developments here and also to discuss an experiment that we should do in Pasadena. I don't know whether you know of the recent Jacob-Monod work and so I will just mention the salient features.

First of all it seems very likely that control mechanisms e.g., repression, may operate at the genetic level. Jacob has found constitutive mutants of the β-galactosidase gene which are different from the *i* mutants and act only in the cis position. In the same region are found mutants which result in the failure to synthesize both β-galactosidase and the permease. If this is so then it means that when the system is repressed the best way to explain this finding is to assume that the templates have to be continually renewed and are destroyed either after one or a small number of synthetic steps. Into this picture we can fit the finding that the continued expression of the β-galactosidase requires the presence of the gene. Our picture is as follows: proteins are synthesized in the ribosomes. Most of the RNA is structural and carries no genetic information. This RNA seems to have a base composition which is constant throughout nature. Into the particles are then fed the RNA templates which have a limited life, and which are direct copies of the DNA genes. The obvious candidate of such template RNA is the Volkin RNA found after T2 infection. This has an A/G ratio which is the same as the phage DNA and is unstable, showing turnover. In order to prove this theory directly we have to investigate the distribution of the Volkin RNA amongst the ribosomes of the cell.[3]

Brenner went on to describe the experiments under consideration, the most urgent of which was

to study by density gradient centrifugation the distribution of the new RNA amongst the particles. I have been looking at the stabilization of particles by formaldehyde to CsCl and it looks as though this works. I propose to clean up the technique and suggest that in Pasadena we do this experiment. I shall be arriving in Pasadena on Monday 6th June and will stay for at least a month, skipping the Gordon Conference. It would help us if you had available all the density labels that we might need, heavy bases, heavy amino acids, etc. You know exactly what we will have to use.

Please write and give me your reactions to these ideas.

Yours
Sydney[3]

By the time Brenner and Jacob arrived in Pasadena in the early summer of 1960, word of the hypothetical new RNA had reached the United States, where the reception was in the main skeptical. Max Delbrück voiced categorical disbelief about the notion of messenger RNA. However, he had something of a reputation for often being wrong in his scientific predictions.[b] Thus, when Brenner heard of this pronouncement, he told Jacob, "We're fine—because Max doesn't believe it!"

The two visiting scientists began their experiments on June 7, 1960, optimistically anticipating that they would be done before the month was out. Meselson offered logistical and technical help before he left Pasadena to propose to his first wife, leaving his colleagues with the prediction that the experiment would be either a total flop, or less probably, a real smash. While in principle relatively straightforward, in practice the work was fraught with technical complications. For instance, Meselson had obtained his supply of heavy isotopes from a group of Russian scientists who had generated the prized material as a special favor to Linus Pauling. Pauling was appreciated for his opposition to American harassment of so-called Communist sympathizers and had been elected to membership in the Russian Academy of Sciences in 1958. Very little of the precious isotopes was available, so little that Brenner and Jacob had to add large amounts of unlabeled carrier material and content themselves with tracking very small quantities of isotope.[4]

Maintaining the integrity of the ribosomes in the very high ionic strength cesium chloride solution was a maddeningly persistent problem. The cesium chloride was about eight molal in concentration, a very strong salt solution. So Brenner went to the literature in search of bacteria in which ribosomes might be more robust. Much to his satisfaction, he discovered that some grew in the Dead Sea. For a while he considered obtaining these, but dropped the idea when he realized that they would be resistant to infection with the bacteriophages they planned to use, not to mention the considerable time and energy required to thoroughly characterize these odd beasts.

The days came and went, and the end of their planned month-long stay in Pasadena loomed ominously. However, before he left Cambridge for Caltech, Brenner had carried out an ingenious proof-of-principle experiment, based on the premise that, when bacteria are starved for magnesium their ribosomes disintegrate, but readily recover if the crippled cells are returned to medium containing normal levels of magnesium. His results showed that, after phage

---

[b] His scientific genius was never in question, but the strong-willed Delbrück could be wrong, with surprising frequency, in his scientific predictions or pronouncements. People who knew him well during his years at Caltech would sometimes joke that if Delbrück believed something to be wrong, it must be right.

infected their bacterial hosts in magnesium-depleted medium, the cells do not recover unless magnesium was added back to the medium. "That said to me that after phage infection you don't make any new ribosomes. So I knew then that we had to be right."

This assurance kept Brenner and Jacob motivated. Still, the experiments at Caltech were nerve racking. They used a lot of precious isotope, and if anything went wrong—if the centrifuge broke down, or a rotor was damaged—they had to start all over again. "Of course we got little sympathy from others at Caltech because we were trying to do an experiment that really required a lot of systematic trouble shooting. People said, 'Well, such experiments normally take a year or so to work out—and you guys want to do them in a few weeks.' Delbrück was particularly unsympathetic because he didn't believe the whole idea to begin with."

François Jacob commented graphically on this dispirited time.

> In a burst of compassion, a biologist by the name of Hildegard [Lamfrom][c] had taken us under her wing and, to give us a change of scene, driven us to a nearby beach. There we were, collapsed on the sand, stranded in the sunlight like beached whales. My head felt empty. Frowning, knitting his heavy eyebrows, with a nasty look, Sydney gazed at the horizon without saying a word. Never yet had I seen Sydney Brenner in such a state. Never seen him silent. On the contrary, he was an indefatigable talker at every opportunity. A tireless storyteller, able to discourse for days and nights on end. Interminable monologues on every conceivable subject. Science, politics, philosophy, literature, anything that cropped up. With stories he made up as he went along. Generously laced with jokes. With nasty cracks, too, at the expense of just about everyone. An excellent actor, he could render a speech in Hungarian, a lecture in Japanese, mimic Stalin or Franco. He went without a break from one register to another.[5]

Brenner was indeed disconsolate—and perplexed. Lazing on the beach that day, his mind churning for explanations as to why the ribosomes were not behaving as expected, he feared that he and Jacob might indeed have to settle into an extended period of systematic (and boring) troubleshooting to determine appropriate experimental conditions. Suddenly, lightning struck. "I abruptly realized that the cesium must be competing with the magnesium required to stabilize ribosomes; not very efficiently, but enough to wreck our

---

[c] Hildegard Lamfrom, a biochemist at Caltech working on the synthesis of hemoglobin in reticulocytes, had considerable experimental experience in handling ribosomes and may have reminded the duo of the stabilizing effect of magnesium on ribosome structure.

experiments. The magnesium we were using was only a thousandth molar, while the cesium was eight molar. I sprang up and shouted to François, 'It's the magnesium! It's the magnesium!' "

The group piled into Hildegard Lamfrom's car and dashed back to the laboratory to set up a new experiment with different concentrations of magnesium, realizing that this might be the last one they performed for quite a while. Jacob was so nervous he spilled some of the radioactive phosphate into a water-bath, prompting them to hide the piece of equipment behind a Coca-Cola machine in the basement. Several years later when Brenner returned to Caltech on another visit, he perversely asked whether he might have a Coca-Cola to drink. " 'Oh yes, of course,' they responded. 'Is the machine still in the basement?' I innocently enquired. I wanted to see if this radioactive water bath was still there. It wasn't!"

Imagine the consternation when, in the middle of this critical experiment, one of their worst fears—that the centrifuge might break down—was realized. The cesium chloride gradients in which the cell lysates were being centrifuged to their equilibrium position were mechanically stable, but they were not thermally stable. So the rotor had to be removed and transported to a cold room while another centrifuge was procured. When this hurdle was successfully overcome and the centrifugation process finally completed, Brenner and Jacob had to carry the centrifuge rotor to the laboratory, being careful not to disturb the contents of the centrifuge tubes. People rushed to their assistance, helpfully clearing a path and making sure that no one suddenly came barging out of a room. In those days misting water into the circulating air cooled the buildings at Caltech, and Brenner was soaking wet by the time he safely reached the laboratory where the experiment was to be continued.

Once the centrifuge tubes were removed from the rotor and secured, their bottoms had to be carefully punctured with a needle so that the contents could be distributed among a rack of test tubes. This was achieved by gently moving the rack by hand, collecting equivalent volumes in each tube by counting drops. If the puncture holes were too small, drops would not emerge. If too big, liquid would escape in a horrible gush. "François was very nervous when I pierced the tubes," Brenner related. "I was too, and when I did the first one I had a bit of a shake at the beginning, so I missed the first four drops. But finally we dispensed with all three tubes without further incident."

On that day the Democratic Party formally nominated John F. Kennedy as its candidate for the presidency of the United States. Everyone in the building was crowded around the television to view this historic event—except

Brenner and Jacob, who parked themselves in the basement where the scintillation counters (machines used to measure the amount of radioactivity in each tube) were kept. With bated breath and eyes firmly glued to the screen of the counter, they watched to see whether the profile of radioactivity demonstrated the association of newly-synthesized radio-labeled RNA exclusively with extant ribosomes—or not.

"The radioactivity slowly began to rise and we became absolutely delirious," Brenner stated. "I shouted in French, 'Ascendez, ascendez, It's rising, it's rising!' Then we realized it was time for the radioactivity to drop if the experiment was correct. So we were both shouting at this machine, 'Go down, go down, down, down.' The counts in the next tube went up a bit, but the increase was less and I excitedly said, 'It's less, it's less!' " After all was said and done, the results of the experiment were unambiguous. A single peak of radioactively labeled RNA coincided precisely with the position of old ribosomes.

After the initial excitement abated, Brenner and Jacob returned to their respective institutions, and Brenner spent several months executing necessary control experiments. In the central experiment at Caltech, the extant ribosomes had not been directly visualized. Their presence had been inferred by the observation of a single band of newly synthesized radioactive RNA at the expected position in the cesium chloride gradients. Now Brenner formally demonstrated that in uninfected cells, in which ribosomes were both density labeled and tagged with radioactivity, the radioactivity sedimented exclusively with preexisting heavy ribosomes. There was no hint of new ribosomal synthesis.[6] However, these experiments held none of the tension and suspense of those carried out at Caltech. The pair was supremely confident that the results obtained in Pasadena were valid well before these controls were completed. Before long "Sydney resumed his interminable monologues and jokes, his explosive outbursts and spiteful cracks. Amazing Sydney. Always playing with ideas, things, words. A mind as agile as his hands."[7]

In the winter of 1960, Brenner wrote a manuscript and submitted it to *Nature* within days of the year's end so that the paper would have a 1960 submission date. He forwarded copies to his coauthors Jacob and to Matt Meselson. The paper, entitled An Unstable Intermediate Carrying Information from Genes to Ribosomes for Protein Synthesis, proposed Brenner and Jacob's original hypothesis. But it included also two straw-man alternatives: that DNA instructed protein synthesis directly or that a small number of new ribosomes were indeed synthesized. The paper discussed the expectations of each of these models and explained how the incorrect hypotheses were experimentally eliminated.

In addition to the results obtained in phage-infected cells, Brenner and Jacob obtained the same result in uninfected bacteria and reasonably concluded that the association of newly synthesized RNA (mRNA) with preexisting ribosomes was a general phenomenon, at least in prokaryotes, not a phage-specific oddity.

Meanwhile, another research group was carrying out similar experiments. Brenner learned this news from a letter by Jacob, who wrote:

> As you have probably heard, François Gros [a young molecular biologist at the Pasteur Institute who was well known to Jacob and Monod] and Jim [Watson] have now found the message RNA by two methods. . . . Apparently the whole US biochemists are now working on this! It is therefore worth not to delay too much our publication.[8]

Despite this news and having submitted their manuscript for publication, Brenner and Jacob saw no particular reason to make direct contact with the Watson laboratory. But when Matt Meselson, now a faculty member at Harvard with Jim Watson, received a copy of the Brenner-Jacob-Meselson manuscript, he hurriedly informed Brenner that Watson and his colleagues (François Gros, the Swiss biochemist, Alfred Tissières, and the physicist-turned-molecular biologist Wally Gilbert) had independently concluded that messenger RNA could be detected in uninfected *Escherichia coli*. In fact, Watson was preparing his own manuscript on the topic. Meselson's concern had little, if anything, to do with winning priority for a major scientific discovery. On the contrary, he was uneasy about alienating Watson, a senior colleague at Harvard and an internationally prominent scientific figure clearly destined to win a Nobel Prize. Up to that point, communication between Brenner and Jacob and the Watson group had been open and courteous. Indeed, on September 21, 1960, a few days after he received Jacob's letter informing him of the ongoing studies in the Watson laboratory, Brenner penned a friendly note to Watson, asking for assistance with his own experiments.

> Dear Jim,
>
> Could you please do me a favour? I understand that someone in your group has an amino acid analysis of the ribosome protein, and I wonder whether you are prepared to send it to me. . . .
>
> Sincerely,
> Sydney.[9]

Having independently discovered messenger RNA, Watson was keen to have his work published at the same time as the manuscript written by Brenner and his team. On February 16, 1961, he contacted Jacob by telegram, who, in turn, immediately relayed a note of concern to Brenner.

> Dear Sydney,
>
> I just received a telegram from Jim which I enclose in this letter.[d] So I hope they will manage their paper soon enough to have them published together in the same issue. This is by far the best solution and we should have done this before.[10]

In addition to agreeing to publish back-to-back manuscripts, Jacob and Meselson suggested that their own paper be modified to explicitly address findings in the Watson laboratory. Brenner agreed. In a letter written on February 21, 1961, he informed Watson of the revisions he made to the Brenner/Jacob/Meselson paper. He also politely indicated that he was holding up publication of the manuscript.

> ... I have written to the Editor of Nature saying that we wish to hold up publication so that simultaneous publication can be achieved; for this reason I would be grateful if you could send me a copy of your paper as you submit it.[11]

Watson was no newcomer to the RNA world. He had long been interested in the molecular mechanism of protein synthesis and the role of RNA in this process. Indeed, soon after the elaboration of the DNA structure with Francis Crick, he and Alex Rich spent an exasperating and ultimately fruitless three years trying to identify a structure for RNA that might be revealing of its function in protein synthesis. In his Nobel Lecture delivered on December 11, 1962, and published soon thereafter, Watson wrote:

> The problem whether RNA was a one- or several-chained structure remained unanswered. We then considered the possibility that RNA might have a regular structure only when combined with protein. ... It thus seemed logical to turn our attention to a study of ribonuclear proteins (ribosomes) since upon their surfaces protein was synthesized. ... Thus when Alfred Tissières and I came to Harvard's Biological Laboratories in 1956, we initiated research on the ribosomes of ... E. coli.[12]

---

[d] Attempts to locate this telegram were regrettably unsuccessful. However, one assumes that its overall tone was not especially friendly.

In this and in several subsequent accounts of his studies on ribosomes, Watson explicitly mentioned the work on messenger RNA initiated by his Harvard research group:

> We were ... convinced that similar messenger RNA would be found in uninfected bacteria. Its demonstration then presented greater problems, because of the simultaneous synthesis of ribosomal and soluble RNA. François Gros had then just arrived to visit our laboratory. Together with Mr. Kurland and Dr. Gilbert, we decided to look for labeled messenger molecules in cells briefly exposed to a radioactive RNA precursor [P^{32}]. [Our] experiments with T2 infected cells suggested that the T2 messenger comprised about 2–4% of the total RNA and most of its molecules had lives less than several minutes. If a similar situation held for uninfected cells, then, for any short interval, most RNA synthesis would be messenger. There would be no significant accumulation since it would be broken down almost as fast as it was made. Again the messenger hypothesis was confirmed.[12]

The two papers announcing the existence of mRNA were published back-to-back in *Nature* in mid-May 1961.[e] In the interim Jacob and Monod had written a lengthy review in the *Journal of Molecular Biology* that appeared just a few weeks later. Obviously privy to the results of the Brenner/Jacob/Meselson experiments well before they were published in *Nature*, Jacob and Monod addressed these in some detail in their review. As the authors had little information about the work by the Harvard group, their only reference to that effort was a sentence stating:

> ... a small fraction of RNA, first observed by Volkin and Astrachan (1957) in phage infected *E. coli*, and recently found to exist also in normal yeasts (Ycas and Vincent, 1960) and coli (Gros et al. 1961) [the Watson study], does seem to meet all the qualifications [for mRNA] listed above.[13]

The *Nature* article by the Watson group, therefore, may not have attracted the attention it might otherwise have enjoyed.

By no means did everyone accept the messenger RNA stories. Immediately after completing the experiments in Pasadena, Brenner presented a seminar on the discovery at Stanford University that was received with some skepticism by Arthur Kornberg. There were other skeptics, among

---

[e] Brenner S, Jacob F, Meselson M. 1961. An unstable intermediate carrying information from genes to ribosomes for protein synthesis. *Nature* **19**: 576–581; Gros F, Hiatt H, Gilbert W, Kurland CG, Risebrough RW, Watson JD. 1961. Unstable ribonucleic acid revealed by pulse labelling of *Escherichia coli*. *Nature* **19**: 581–585.

them Henry Harris, the newly appointed head of the Department of Cell Biology at the John Innes Institute in Norwich, who argued against the existence of mRNA (although he correctly predicted that mRNA in eukaryotic cells is not short-lived).[14] In particular, many biochemists pressed for more direct evidence; data showing that if one incubated a selected mRNA fraction with suitably prepared cell-free extracts, one could observe synthesis of the cognate polypeptide. "It took a lot of time to convince people that you could actually do an experiment like this and take it as evidence," Brenner related.

Jacob thought it marvelous that the notion of messenger RNA was emphatically reinforced by his β-galactosidase experiments and that its discovery explained gene regulation—at least in bacteria. Later he and Brenner demonstrated that galactokinase messenger RNA is produced only when the galactokinase operon is induced with an analogue of β-galactosidase. Indeed, the first formal proof that induction of gene expression involves new messenger RNA synthesis is in a paper published by Brenner and Jacob in the French journal *Comptes Rendues.*[f]

Max Perutz wasted little time in exploiting the discovery to educate MRC Secretary Harold Himsworth as to the imperative of protecting the LMB faculty from prospectors in America.

> I am writing to tell you about a major advance made in this Unit by Brenner, in collaboration with F. Jacob at the Pasteur Institute and M. Meselson at the California Institute of Technology. . . .
>
> Members of the Unit continue to be pressed to accept posts in the United States. Brenner has been offered the directorship of the Carnegie Institute of Genetics at Cold Spring Harbor, Kendrew the chair of biophysics at Ann Arbor and another one at Los Angeles, and Crick the directorship of a new laboratory of Molecular Biology at the NIH. Luckily they have all turned deaf ears to the Siren's calls.[15]

"Nineteen sixty-one was a very exciting year," Brenner concluded. "During that year we published two papers. One was on the messenger RNA experiments and the other was on the genetic code." General Nature of the Genetic Code for Proteins, the title of the latter paper published by Crick, Barnett, Brenner, and Watts-Tobin in the final issue of *Nature* in 1961, is a classic in molecular biology—perhaps *the* classic paper published during the so-called golden age of molecular biology.

---

[f] Jacob F, Brenner S. 1963. Genetique Physiologique—sur la regulation de la synthese du DNA chez les bacteries: l'hypothese du replicon. *CR Acad Sci* **256:** 298–300.

13

A Triplet Genetic Code

One of the most aesthetically elegant experiences of my life

S EVERAL YEARS BEFORE THE DISCOVERY OF MESSENGER RNA, soluble or transfer RNA (tRNA) was revealed to play a key role in protein synthesis. As we saw in Chapter 6, Francis Crick predicted the existence of this entity as early as 1955 in his theoretical paper On Degenerate Templates and the Adaptor Hypothesis, written for the RNA Tie Club. This treatise (which was never formally published) has been hailed by some as the finest example of theoretical biology in the 20th century and by Crick himself as his "most influential unpublished paper."[1] Not much later the Americans Paul Zamecnik and Mahlon Hoagland identified Crick's adaptors in the course of their studies on protein synthesis. They called this nucleic acid "soluble RNA," but by the early 1960s the more generally used term "transfer RNA" was adopted.

The essential mechanism of decoding genetic information and translating it to specific amino acids was now fully emerging: information for the assembly of a particular polypeptide encoded in DNA is first transcribed into messenger RNA. Once assembled at ribosomes, nucleotides in the messenger pair with complementary partners of individual transfer RNAs (each of which is charged with a cognate amino acid), resulting in the incorporation of amino acids in a specified order, to generate a unique polypeptide chain.

The stage was now set to decipher the genetic code, but crucial questions remained. How many nucleotides encode a single amino acid? What is the actual nucleotide code for each amino acid? Is the code script punctuated? What signals the beginning and end of the code in a gene? Perhaps most perplexing, why are there 64 possible triplet codons for only 20 amino acids?

As early as the mid-1950s, another RNA Tie Club communication, by Crick, Leslie Orgel, and John Griffith, presented the notion that the code was comma-free, that is, functional groups of nucleotides (codons) are not punctuated.[a] Furthermore, as we saw in Chapter 7, Brenner had demonstrated the theoretical impossibility of an overlapping genetic code. However, by the early 1960s, progress with the so-called coding problem had been fitful and could be justly characterized as more conjectural than experimental.

Faced with these challenges Brenner and Crick revisited the issue. But the time for theorizing was over; the two now sought direct experimental evidence of how the code is read. As Crick candidly put it, "The time is rapidly approaching when the serious problem will be not whether, say, UUC is likely to stand for serine, but what evidence can we accept that establishes this beyond doubt."[2] During the course of a relatively brief period, in the early to mid-1960s, Brenner, Crick, and their respective colleagues—sometimes publishing together, sometimes independently—established the triplet nature of the genetic code. As we shall see in the next chapter, they also defined the distinction between "sense" and "nonsense" in the code and extended the notion of nonsense to explain polypeptide chain termination during normal protein synthesis. Brenner even identified some of the nucleotides that specify nonsense in the code by pure genetics—long before DNA sequencing was possible. Ultimately, he succeeded in his passionate quest to demonstrate colinearity between a gene and its polypeptide product, in a most unexpected and ingenious fashion.

These penetrating and rewarding contributions to molecular biology were wrought from the simple bacteriophage system by Brenner and Crick in the period between 1961 and 1965, a system requiring little more than Petri dishes, agar, pipettes, a few incubators—and two formidable scientific intellects. These efforts alone ought to have merited a Nobel Prize for Brenner and a second Nobel for Crick. During the decade of the sixties, many molecular biologists were honored as Nobel Laureates,[b] but these did not include Brenner, an oversight perplexing to many, including Sydney.

---

[a] This theoretical study, entitled "Codes Without Commas," was formally published under the same title in 1957 (Crick FC, Griffith JS, Orgel LE. 1957. Codes without commas. *PNAS* **43:** 416–421). The authors presented an elegant theoretical analysis that led to the conclusion that only 20 codons specify sense, i.e., amino acids. The remainders are nonsense. Horace Judson referred to this contribution as "the most elegant biological theory ever to be proposed and proved wrong" (Judson, p. 315).

[b] The Nobel Laureates recognized for their contributions to the so-called golden age of molecular biology were James Watson, Francis Crick, Max Perutz, Maurice Wilkins (1962), François Jacob, André Lwoff, Jacques Monod (1965), Robert Holley, Gobind Khorana, Marshall Nirenberg (1968), Max Delbrück, Alfred Hershey, and Salvador Luria (1969).

It is difficult, if not impossible, to sort out the specific contributions of Crick and Brenner to their remarkable partnership. The pair enjoyed an intellectual complementarity that elevated their professional relationship to historic proportions. Crick liked to pursue ideas and hypotheses to their experimental conclusion, but he needed someone like Brenner to challenge him with new ideas. Brenner, on the other hand, required someone like Crick to filter his constant flow of ideas, many completely undeveloped at their inception. He was also less regimented than Crick and benefited greatly from the latter's discipline in bringing things to completion. "I think that had Francis Crick not existed I might have never written a paper in my life," Brenner stated. "It was Francis who made me write papers; because once I had solved a problem I lost interest in it. But Francis used to lock me in a room and say: 'You've got to write it up.' "[3]

Both men could talk endlessly and enthusiastically, and both loved to think aloud, preferably in the company of quick-minded, critical, and attentive listeners, such as each other. In Brenner's words: "Most of these conversations were just complete nonsense. But every now and then a half-formed idea would be taken up by the other one and really refined. I think a lot of the good things we produced came from these completely mad sessions."[4] Conversation, discussion, criticism, and argument were fundamental to their relationship, and they shared an office even when space in the new Laboratory of Molecular Biology obviated the need.

Brenner quickly discovered that Crick could be a very severe audience, one who challenged his colleague on poorly articulated ideas or suggestions. "One didn't get away with anything. He asked very penetrating questions and one had to be thinking all the time. But the clarification that came from this sort of dialog was very important." On a nearly daily basis, the pair spent hours exploring all manner of notions and ideas. Both scientists tended to be visual in their thinking and profited from drawing on the blackboards, especially to obtain a sense of the relative size and complexity of cellular and subcellular entities. "Francis was brilliant at visualizing molecular structure," Brenner related.

> He thought geometrically, like I do, not algebraically. Neither of us would sit down and write axioms and them proceed to deduce answers. We used diagrams a lot. Francis was very good at that too. But we were always careful to keep the scale of things in mind. That is very important. You see a lot of cartoons of a bacterium with a little circle inside to indicate the genome. But it's important to realize that there's a millimeter of DNA in that tiny bacterium! So Francis and I tried very hard to stay imprisoned in the physical context of everything.

Believing as he does that informed scientists sometimes unwittingly—sometimes wittingly—cultivate biases and prejudices about unsolved problems, Brenner likes to talk to intellectually stimulating scientists from other disciplines.

> I believe that people who come to a field from the outside, who have not been entrained to the standard approach, can sometimes see things from a different perspective. Gamow didn't know anything about molecular structure, but he saw things from the perspective of a physicist and he could pose problems in a form that no biochemist would or could.

Crick echoed these sentiments: "It was a blissful period because the problems were important,"[5] he wrote in a tribute to Brenner on the occasion of his colleague's 75th birthday.

> Only a few people (most of them friends) were working on them then and, thanks to the Medical Research Council's support, we didn't have to write grant requests and could study whatever we liked. Sydney and I had discussions almost every working day—using several large blackboards—but he also spent long hours in the lab and considerable time reading the literature.[c] He was much better than I at thinking up novel experiments. My role was more that of a critic and clarifier.[5]

Science writer Matt Ridley also documented the Crick-Brenner dynamic.

> The dialogue between Brenner and Crick was a conversation that developed its own rules. There was no shame in floating a stupid idea; but no umbrage was to be taken if the other person said it was stupid. Anyone else from the lab could walk in and interrupt if the door was open, but strangers were directed to see the secretary. Like Watson, Brenner knew a lot more biology than Crick. [On the other hand] Brenner found Crick an "incredible cross-examiner" who always challenged him on how to test an idea with a real experiment.[6]

Crick described his memory to his biographer Robert Olby as fallible, but offered the opinion that "Brenner has an amazing memory . . . he is a traveling encyclopedia on a cornucopia of subjects from medieval history to paleontology and computer science."[7] Crick further volunteered that his collaboration

---

[c] Both Benner and Crick read the literature voraciously. Matt Ridley commented: "Crick . . . was a ravenous consumer of others' results, from even the most obscure publications, and he had formidable powers of concentration. When Aaron Klug once asked why Crick was wasting time on an obviously useless paper the response was, '[T]here might be a clue in it' " (Ridley M. 2006. *Francis Crick: Discoverer of the Genetic Code*, p. 103. Eminent Lives Series, London).

with Brenner was not only fundamental to the development of ideas, "but it was all such fun. It says much for his tolerance and good temper that there was never an angry word between us. Happy days!"[8]

Crick had largely confined himself to a theoretical role in the hectic scramble to investigate the nature of the genetic code and its operation. However, he soon began carrying out experiments on phage mutagenesis with his own hands. He was then keenly interested in the problem of mutational suppression (sometimes referred to as mutational curing), a phenomenon in which a mutant phenotype is eliminated in the presence of a second mutation elsewhere in the genome. A particularly challenging example of mutational suppression surfaced from experiments carried out by Alice Orgel (Leslie Orgel's wife and a graduate student under Brenner's supervision). She demonstrated that the polycyclic aromatic dye proflavine, a known mutagen, cured mutations generated by the same compound, but had no effect on mutations induced by other mutagens, such as bromouracil.

Crick and Brenner tossed the notion of suppressor mutations back and forth. The pair reasonably hypothesized that if a single mutation affected the function of a protein, a second mutation in another gene that affected a different protein might compensate or suppress the phenotype of the first mutant. But they were disturbed by the huge number of suppressor mutations that Crick observed when phage was exposed to proflavine—more than could be comfortably accommodated by phenotypic suppression. Furthermore, Crick noted that, in every case, a proflavine-induced suppressor mutation mapped very close to the mutation it suppressed, suggesting that this mechanism was local rather than one acting at a distance.

As was Watson and Crick's tendency some years earlier, Brenner and Crick frequented The Eagle, a local pub well populated with Cambridge University types. While at the pub one Saturday morning, Brenner had another remarkable epiphany—no less sudden and providential than the messenger RNA insight that had emerged shortly before in his rooms at King's College.

> I suddenly had the notion that if, as people had suggested, the planar dye proflavine inserted itself between base pairs, one might have a situation where the DNA "thought" that the dye molecule was another base and the cell stuck in an extra base on the other strand during DNA replication, or made a compensating deletion during replication. So the idea emerged of a connection between mutations and base additions and deletions.

This notion was supported by the suggestion by Crick, Orgel, and John Griffith in 1955 that one could write commaless codes in which one reading frame of nucleotide triplets made sense while in every other frame it was nonsense.

The following Monday Brenner and Crick began experiments to test this hypothesis. They showed that all known spontaneous phage host-range mutations could be reverted with base analogues such as the thymine analogue bromouracil. However, the great majority of spontaneous mutations in the *rII* gene were not revertible with base analogues. Furthermore, mutations generated by base analogues (such as bromouracil) were not observed when proflavine was used as the mutagen. Crick and Brenner produced another theoretical paper entitled The Theory of Mutagenesis,[d] another classic in theoretical biology. Here they proposed that, in addition to the well-known nucleotide substitution mutations—transitions and transversions—there was another category—addition and deletion mutations. *"Acridines act as mutagens because they cause the insertion or the deletion of a base pair,"* the authors noted (italics in original).

"One could start with a mutant arbitrarily called 'minus,' which was due to the loss of a base," Brenner explained. "All the mutations that suppressed this minus, such as those caused by proflavine, would be 'pluses' such that when you added a single 'plus' to a single 'minus' they would cancel out and the phenotype would no longer be mutant." Brenner and Crick soon realized that if they could prove this model they would be able to determine whether the code was, indeed, spelled out in sets of three nucleotides—as everyone had long believed. "All we had to do was to ask for any mutation that was revertible by proflavine, how many bases must be added or missing for the mutation to be suppressed, i.e., to restore the normal reading frame."

Brenner speaks nostalgically of these experiments as "a sort of apotheosis of a genetic analysis." Both he and Crick marveled at the fact that fundamental conclusions emerged from such technically simple experiments. Literally dozens of experiments could be carried out more or less simultaneously, with results available in a day or so. The observations were simply to score whether or not growth of the phage occurred. "From this pattern it seems mad that you could deduce the actual triplet nature of the genetic code." The English microbial geneticist William Hayes later referred to these results as "a masterpiece of genetic analysis."[9]

Crick carried out many of the plus and minus experiments with his own hands. Crick's efforts exasperated Muriel Wigby, an experienced technician

---

[d] Brenner S, Barnett L, Crick FHC, Orgel A. 1961. The theory of mutagenesis (editorial letter). *J Mol Biol* 3: 121–124.

who joined the laboratory soon after Brenner arrived in Cambridge in 1957. She remained his assistant for over 30 years, breeding worms (see Chapter 15) and bacteriophage alongside a slew of LMB scientists who would become Nobel Prize winners. Whatever her opinions about Crick's intellect, Wigby was not impressed by Crick's technical abilities, accusing the man of being "terribly clumsy."[10]

Wigby worked beside Brenner and Crick long enough to comprehend their whims and foibles in the laboratory. She found Crick generally more difficult than Brenner, particularly his inclination to blame her when experiments failed.[e] "When Francis gave me an experiment to do and it didn't come out the way he expected he would insist that there must be something wrong with the way I had done it," she protested.[10] In contrast, Brenner rarely questioned her technical competence. "But when experiments didn't work and he became angry or frustrated, he would walk back and forth in the lab muttering and sometimes swearing—in Afrikaans."[10]

Brenner noted:

> An interesting thing about these experiments was that it was a real house of cards theory. You had to buy everything. You couldn't take one fact and let it stand by itself and say the rest could go. Everything was so interlocked. You had to buy the plus and minuses and you had to buy the triplet phase; all these went together. It was the whole that explained it and if you attacked any one part of it the entire thing fell apart. So it was an all or nothing theory. And it was very hard to communicate to people. However, this was one of the most beautiful, aesthetically elegant experiences of my life, in which, just by doing these little operations you landed up with a detailed description of the molecular structure of living matter.

Brenner and Crick published their observations on the triplet nature of the code in a landmark paper entitled General Nature of the Genetic Code for Proteins that laid bare some fundamental features of the triplet genetic code.[11]

---

[e] Crick's impatience (and sometimes incompetence) with wet bench research was also commented on by the biochemist Mahlon Hoagland, who worked with him on tRNA at the Cavendish. "We would do an experiment and get some variation in results that Francis felt obliged to analyze and ponder at length. I would assure him that the variations were very likely an error—we would not find them if we repeated the experiment. ... He had an uncanny ability to analyze and criticize, in *detail*, the experiments of others, but at the bench he became mired in the day-to-day messiness and inconclusiveness." (Olby R. 2009. *Francis Crick: Hunter of life's secrets*, p. 267. Cold Spring Harbor University Press, Cold Spring Harbor, New York.)

A group of three bases (or less likely, a multiple of three bases) codes one amino acid.

The code is not of the overlapping type. . . .

The sequence of the bases is read from a fixed starting point. This determines how the long sequences of bases are to be correctly read off as triplets. There are no special "commas" to show how to select the right triplets. If the starting point is displaced by one base, then the reading into triplets is displaced, and this becomes incorrect.

The code is probably "degenerate"; that is, in general, one particular amino acid can be coded by one of several triplets of bases.[11]

General Nature of the Genetic Code for Proteins was rich in conclusions and conjecture but contained little of the mass of experimental data on which they were based. In fact, the complete experimental details of these studies were not published until 1967.[12] While thumbing through the massive final draft of the paper that would occupy 73 pages of the *Philosophical Transactions of the Royal Society*, Crick idly commented to Brenner that the two of them were likely the only individuals in the world who would read the published paper. They, therefore, plotted to insert within the manuscript a bogus literature citation, credited to a figure of historic prominence who had absolutely nothing to do with biology. After some discussion, the pair settled on referencing a personal communication from Leonardo da Vinci. Crick commented: "[O]ne (unknown) referee passed it without comment, but we had a phone call from the other referee, who asked, 'Who's this young Italian working in your lab?' So reluctantly we had to take it out."[13]

Not all the multiple suppression experiments obeyed the simple plus or minus rule. Rare exceptions emerged in the data set. Many scientists might dismiss these as fundamentally unimportant curiosities that a reluctant graduate student may sometime wish to explore; not Brenner and Crick. "For a long time we hung on to the 'don't worry hypothesis'—that sooner or later there'll be an explanation for them." About five years later, explanations derived from their own experimental observations did, indeed, emerge.

14

Deciphering the Triplet Code

I did sequence three bases in DNA—by genetics alone

Iᴛ ᴡᴀѕ ɴᴏᴡ ᴏʙᴠɪᴏᴜѕ ᴛᴏ ᴍᴀɴy, especially among the community of biochemists, that it should be possible to play any "tape" on any "machine" simply by adding purified messenger RNA to a preparation of ribosomes in vitro and observing the synthesis of the appropriate polypeptide. Crick and Brenner initiated experiments in which they added turnip yellow mosaic virus mRNA to preparations of *Escherichia coli* ribosomes. At that time it was unknown that the requirements for protein synthesis in bacteria and mammalian cells differ, and so these experiments came to naught. Soon afterward, the elucidation of which triplet codons specify which amino acids emerged from experiments performed in the laboratories of Severo Ochoa at New York University and the late Marshall Nirenberg at the National Institutes of Health. When Nirenberg carried out experiments with a synthetic RNA polymer (polyuridylic acid), using a synthetic deoxyribonucleotide polymer as a negative control, he observed that poly(U) instructed incorporation of the amino acid phenylalanine into polypeptide chains.[a]

The success of the biochemical approach adopted by the American investigators alerted Brenner and Crick to the limited development of modern biochemistry in the United Kingdom. In contrast, trainees emerging from the laboratories of investigators such as Ochoa, Nirenberg, and Arthur Kornberg had extensive biochemical expertise and a huge armory of reagents. Regardless, Brenner is not hesitant to voice his bias in favor of genetics (see p. 108). "I have to admit that one of the things that slowed us down was that, conceptually, doing things by biochemistry seemed like awfully brute force. Francis

---

[a] Nirenberg shared the 1968 Nobel Prize in Physiology or Medicine with Gobind Khorana and Robert Holley. Nirenberg passed away in early 2010.

and I had come to the position of prizing experimental ingenuity beyond anything else; of being elegant in the sense of being able to toss out a genetic experiment and learn something profound from it."

Brenner, therefore, continued to exploit the power of bacteriophage genetics in ingenious ways. One of these genetic triumphs led to the deciphering of triplet codons that signal STOP during normal protein synthesis, thereby terminating a growing polypeptide chain. These experiments, in turn, opened a backdoor to proving the colinearity issue that had eluded him and Benzer a few years earlier.

Seymour Benzer had shown that he could suppress particular mutations in the bacteriophage *rII* gene by generating certain secondary mutations. Brenner and Crick suspected that these suppressible mutations were nonsense—mutations within codons that did not specify known amino acids, but were cured by mutations that specifically corrected the nonsense mutations. In the early 1960s Richard Epstein and Charley Steinberg identified a nonsense suppressor gene in *E. coli* that was famously dubbed the "amber" suppressor.[b] Thus when Brenner and his colleague Jonathan Beckwith identified a second suppressor gene in 1965, they playfully called it "ochre." "I thought for a moment that I would call them *umber* mutants, to go with *amber*," Brenner stated. "But I realized this wouldn't go down in England because there are parts of England where 'amber' is 'umber' and 'umber' is 'oomber.' So there'd be a lot of confusion." Subsequently, a third kind of nonsense suppressor called "opal" was identified in the Brenner laboratory.[c]

In the midst of these studies, it occurred to Brenner that nonsense codons may have originally evolved to terminate elongation of polypeptide chains during normal protein synthesis. By that logic, a suppressor might be simply a mutation that shifts the triplet codon reading frame, so that the nonsense codon, no longer in frame, becomes sense. Brenner was aware that the

---

[b] The following is loosely paraphrased from a letter that the phage geneticist Dick Epstein wrote to Frank Stahl around that time. "We also managed to convince Harris Bernstein (then a graduate student working on *Neurospora* genetics) to help, and offered him the dubious reward of naming the mutants after him. Harris had the nickname Immer Wieder Bernstein (Forever Amber in German). That night we isolated several of the desired mutants and named them 'amber mutants.'" See http://www.sci.sdsu.edu/~smaloy/MicrobialGenetics/topics/rev-sup/amber-name.html.

[c] No one seems to recall who first used the word "opal" for nonsense mutation. In response to a letter from the Assistant Editor for Science of the Oxford English Dictionary in 1988, Brenner wrote: "Some called the third [nonsense] triplet opal but it never really caught on (Letter to Mr. P. M. Gilliver, 22 November 1988).

bacteriophage head protein constitutes a major fraction of the total phage protein synthesized in bacteria (see Chapter 11). If his hypothesis was correct, distributing random amber (nonsense) mutations throughout the gene encoding the phage head protein should yield random polypeptide fragments of different sizes. If the experiment worked, Brenner might be able to prove colinearity between nucleotides in the genome and amino acids in its cognate polypeptide, "and we wouldn't have to do any protein sequencing!"

The experiment worked beautifully. A collection of the phage head protein genes, each containing an individual amber mutation that mapped closer and closer toward the end of the gene, produced progressively longer polypeptide chains. In the introduction to the article in *Nature* that documented these results, Brenner wrote:

> It has always been assumed that a simple congruence exists between [the amino acid sequence of a protein and the nucleotide sequence of the gene]. There is no direct evidence for this co-linearity.... In this article we show that a class of suppressible mutations affecting the head protein of bacteriophage T4 produce fragments of the polypeptide chain. This property allows us to prove that the gene is colinear with the polypeptide chain.[1]

Finally, the long-sought goal of proving colinearity was realized. However, shortly before Brenner's discovery, Charles Yanofsky and his colleagues at Stanford University reported colinearity of a segment of the *trpA* gene of *E. coli* and a segment of TrpA protein, at the 1963 Cold Spring Harbor Symposium.[2] In early 1964 the two research teams published their definitive findings within weeks of one another. The Yanofsky paper cordially included a note added in proof stating: "Dr. Francis Crick has recently forwarded a manuscript which deals with a study of colinearity in another system (Sarabhai et al., *Nature*, in press)."[3] Reciprocally, the paper from the Brenner group announced: "Dr. C. Yanofsky of Stanford University has informed us that he has shown co-linearity in an examination of the tryptophan synthetase of *E. coli.*"[4]

Brenner, however, had not finished mining the intricacies of phage genetics. "What still remained very interesting to me was whether we could actually decipher the nucleotide triplets for nonsense mutations. So we performed an experiment that I'm extremely proud of." Brenner knew he could direct a mutation involving a change from C to U or from G to A, based on a judicious choice of mutagens. Furthermore, he realized that, if he could additionally establish that a given mutation affected the coding strand (as opposed to the opposite or noncoding strand), he should be able to decipher the nucleotide composition of the affected triplet. He therefore exposed phages to selected chemical mutagens (based on their ability to cause particular base substitutions), and quantitated and mapped mutations in the head protein gene of the mutant phage that grew on a restrictive (normal) strain, and, in independent experiments, on a nonrestrictive (suppressor) strain. Mutant phage that failed to grow on the restrictive strain carried mutations in the coding strand. Hence, Brenner could deduce what nucleotide substitutions must have transpired in a coding triplet, although obviously he could not predict their order. Based on these observations, Brenner correctly concluded that the most likely nucleotide sequences for the three nonsense codons were UAG (amber), UAA (ochre), and UGA (opal). "It involved an enormous amount of work, but in the end I sequenced three bases in DNA—by genetics alone! Had we gone on plodding in this way we might have been able to work out the entire code." "Genetics is akin to chess for Sydney," one of his colleagues commented. "When handling bacteria and phage he can visualize ten genetic crosses ahead."[5]

The simplest explanation for the molecular mechanism of nonsense suppression was that it involved mutations in tRNA genes. For example, introducing a mutation into the gene for a tRNA that normally read the normal codon UAC could render it capable of reading UAG, a nonsense codon, instead. Brenner, together with John Smith, demonstrated that, in a known suppressor strain, the gene encoding a tRNA had acquired a base change such that its anti-codon now read a STOP codon as sense, and inserted an amino acid.

This chapter of Brenner's career would be incomplete without mention of two other intellectual collaborations. One involved the Frenchman François Jacob, with whom he had forged a close professional and personal relationship. The other involved the immunologist César Milstein, a colleague at the LMB.

A remarkable individual and the author of an intriguing autobiography (discussed in Chapter 11), Jacob began studying medicine with the intention of becoming a surgeon. In June 1940 he interrupted his second year of medical school to join the Free French Forces in London. Jacob was wounded twice, once in Tunisia and again, more seriously, during the Normandy invasion, an event that hospitalized him intermittently for seven months. He was awarded the Croix de la Libération, the highest French military decoration of World War II. After the war he completed his medical studies. His wartime injuries thwarted his ambition to practice surgery, and he turned to biology, obtaining a doctorate in science at the Sorbonne in 1954 under the tutelage of André Lwoff and Elie Wollman.

In the mid-1950s Jacob and Wollman examined the prophage state that occurs when the phage genome is integrated and lies dormant within the bacterial genome. This collaboration led to important insights about the mechanism of bacterial conjugation (sex in bacteria), in which a copy of the replicated chromosome of one bacterial partner is transferred to the other. Jacob's thoughts increasingly turned to understanding the biological control of DNA replication and how such regulation ensures proper segregation of daughter DNA molecules during conjugation. He had little difficulty in engaging Brenner's interest in the sex life of bacteria. A brisk correspondence between Cambridge and Paris ensued, many letters and post cards bearing the salutation "Dear co-sex maniac!"

At that time DNA replication studies were largely in the hands of biochemists, none of whom had any inkling of its complexity. Brenner noted, "You had an enzyme called DNA polymerase, which Arthur Kornberg had discovered and worked on, and you had DNA. And you gave it some substrates and you had the equation: DNA + DNA polymerase + four triphosphates = more DNA. That was the end of the story as far as the biochemists were concerned." But Jacob and Brenner wanted to know more. They wanted to know whether bacteria initiate DNA replication from many sites or from a single unique site and how the event is regulated. Further, they wanted to understand how replicated DNA molecules segregate and how a copy is given to a daughter cell during conjugation.

An opportunity to address these questions arose in the late summer of 1964, when Brenner, Jacob, and their respective families took a vacation together. In mid-1962, Brenner was anticipating returning to the United States the following spring for an extended stay with his family. However, toward the end of August of that year, he informed Watson of the logistical complexities of executing such plans—and his consequent intention to go to Paris instead:

> We now find that it will not be possible to take Jonathan with us since he cannot be taken out of school at all; in addition we must be here in March so that Belinda can write an entry examination for her school. I think it would be best if we called it off and I am sorry to do a Benzer on you [referring to the summer of 1956 when Benzer had to depart Cold Spring Harbor leaving Brenner to handle communicating his work on the phage *rII* gene]. However, children are children. I am determined to get away next year and we have decided to go to Paris. We shall leave Jonathan here and we will be close enough for him to come and spend his half-term break with us.[6]

The two families vacationed at La Tranche sur Mer, a seaside resort in the south of Vendée. Brenner and Jacob spent hours discussing the complexities of DNA replication in living bacteria, elaborating the concept of the replicon, a hypothetical site in the bacterial chromosome where replication always begins. They explored ways of testing this notion experimentally, often using the beach as a surrogate blackboard—pleading with their children not to erase their efforts written in the sand. The two mainly contemplated genetic approaches, ultimately deciding that, as DNA replication is an essential function, they would isolate temperature-sensitive mutants (mutant strains that died at elevated temperatures) and determine which were specifically defective in DNA replication. The large number of mutants that Brenner and Jacob subsequently identified (their alphabetical designation went all the way to Z) confirmed that replication was a complicated process, involving far more than one gene encoding a single DNA polymerase.

The isolation of so many mutants in DNA replication was instructional to Brenner, who realized the power of conditional mutants (mutants that only manifest a mutant phenotype under certain conditions, such as elevated temperature) for studying essential biological functions. As related in the next chapter, this strategy was to be gainfully exploited in his genetic dissection of the worm *C. elegans*.

Brenner's second collaboration involved a brief but important theoretical partnership with César Milstein. In 1984, Milstein added to the steady accumulation of Nobel Prizes at the LMB for his development of the hybridoma technique for production of monoclonal antibodies. Milstein was one of many who pondered the mechanism by which B-lymphocytes in the immune system are able to generate a profusion of antibodies in response to antigenic stimulation. In 1959 the Stanford geneticist Joshua Lederberg suggested that the answer to this question might lie in the unique ability of B cells to generate frequent mutations in their genome spontaneously, a process referred to as "somatic hypermutation."[7]

Being fundamentally interested in mutagenesis, Brenner was quite taken by this hypothesis and wondered how mutations in the genome of somatic cells might arise with such astonishing frequency. In 1966 he and Milstein published a brief theoretical paper in *Nature* (a collaborative effort apparently suggested by Francis Crick) in which they proposed a mechanism that restricted mutations to specific regions of immunoglobulin genes. The Brenner–Milstein model proposed that immunoglobulin genes suffer localized DNA strand breaks, the repair of which is highly error-prone so that the process produces mutations. This hypothesis did not explain primary antibody diversity. Nonetheless, it did become a leading model for the diversification of antibodies in immune responses, the main subject of Milstein's research for the latter part of his life. In fact, a mutagenic process akin to that which he and Milstein proposed has been shown in the past decade to exist in B cells experimentally.[d]

By the early 1960s Brenner had become a household name in molecular biological and genetic circles. Predictably, overtures to relocate his laboratory elsewhere flowed into his office with increasing regularity. Upon learning of the disquieting rumor that Brenner was considering a move to California, Crick wrote:

> I do hope this rumour is out of date. It would be a disaster if you left us, and as I am sure Max [Perutz] has told you, we are going to make quite sure that the MRC gives you all the facilities, in space, money and man-power, that you need. However even the MRC cannot change the Cambridge climate.[8]

The 35-year-old Sydney Brenner was no longer Francis Crick's junior colleague. He was now an equal partner in guiding the future of research at the LMB.

---

[d] See the review article Seki M, Gearhart PJ, Wood RD. 2005. DNA polymerases and somatic hypermutation of immunoglobulin genes. *EMBO Reports* **12:** 1143–1148.

PART 4

Complex Organisms

15

C. elegans

I decided that this was the organism I would go for

IN THE EARLY 1960S CRICK AND BRENNER INITIATED A SERIES of long conversations about what frontiers to tackle next. In their view, the main outlines of the central problems in molecular biology—the tree of life as it were—had been revealed, and the challenge that remained was to populate the various branches. They were convinced that others would fill in this picture with a vengeance, so they spent a lot of time thinking and talking about other problems in biology. Brenner compiled a list of potential research projects that he deemed worthy of consideration. One of these, a project of international proportions, was methodically laid out under the title The Complete Solution of *E. coli* and presented to Crick for consideration and comment.

> The major reasons for wanting to have the "complete solution" of a bacterial cell are
>
> 1. Much remains to be cleared up both in the molecular biology of bacteria (that is, the synthesis of protein and nucleic acid, and the chemistry of genetics, etc.)
>
> 2. Bacteria are likely to prove useful for several relatively unexplored aspects of molecular biology (e.g., membranes, cell division, etc.)
>
> 3. The study of the functioning of the cell as a whole is likely to develop. This would include the various control mechanisms and their rather intricate relationship, and the "cost-accountancy" of the cell considered as a self-replicating chemical factory.
>
> By "complete" one means complete in the intellectual sense, implying that nothing appears to remain which further experiment could not easily explain using well-established facts and ideas.

It is clear that if the cell is going to be considered as a well-integrated chemical factory, information from many different laboratories will have to be pooled. This might argue for an International Laboratory to act as a focus for such work, but there are additional reasons of a technical nature which make the case even stronger.[1]

But as Brenner and Crick discussed these notions that relied on using bacterial cells and phages as model systems, they realized that they really wanted to set their sights on more ambitious goals.

By 1962 there were no longer evangelists bringing a new message. Now the church was admitting everybody and everybody was becoming converted. Being an early Christian must have been exciting. But to become one later is boring because everybody's converted. The only value of being an old Jesuit in the church is that you know what's wrong with the church basically; which the new converts don't. But it's important to keep quiet about this in the early stages of building up the church.

Commenting on the strategic decisions facing Brenner and Crick then, science writer Andrew Brown likened the situation to "the difference between learning to read from a newspaper [at the age of four] and just buying another paper,"[2] hastening to add that Brenner "did not want to spend the rest of his life reading fresh editions of the same newspaper."[3]

Restless about his scientific future and uncertain about that of the LMB, Brenner now considered briefly an offer from Peter Medawar, the illustrious British immunologist recently appointed director of the National Institute for Medical Research at Mill Hill, near London. Medawar was intent on establishing a premier program in genetics and developmental biology. He vigorously pursued Brenner. Brenner was tempted, but then told Medawar that he had decided to remain at the LMB. He also informed Max Perutz in writing that he had not accepted the offer of a position at Mill Hill, but would stay in Cambridge. He pointedly used this opportunity to press the LMB director for change. "I would like to say that my decision to stay in Cambridge will be conditional on future development of our laboratory. . . . As you know, I feel that our laboratory should invest heavily in biology in the future."[4]

This letter bears early testimony to Brenner's wish that the LMB expand its horizons beyond the structure-centric orientation that had won the Cavendish Laboratories so much glory in the past. Having successfully exploited simple biological systems to reveal fundamental aspects of life, Brenner now set his sights on a more ambitious plan—understanding the fundamental aspects of life in more complex organisms. Not everyone at the LMB, however, shared this vision. This brewing protest was reinforced by the passionate international controversy surrounding the use of recombinant DNA technology that was soon to complicate research initiatives in molecular biology (as we shall see in Chapter 18).

Brenner had a long-standing interest in animal cells, and Crick had had some exposure to mammalian cell biology as a graduate student at the Strangeways Laboratory in Cambridge before joining the Cavendish. Like many ambitious molecular biologists of that era, both were especially keen to solve the mysteries of the nervous system. Brenner fully appreciated that, to understand the workings of higher cells, one had to take on the daunting challenges of understanding development and cellular differentiation. "It may be true to say that development was simply a matter of turning genes on and off *al a* Jacob and Monod," he argued. However, that did not help one to understand these processes. "Because at the end of the day what one really wanted was to understand development and differentiation well enough to build a *gedanken* mouse[a]—a functional mouse that can be constructed entirely from knowing the DNA sequence of all its genes."

In the extreme view, one would like to assemble an entire organism from its genome.

> One can look at the head of a virus and see that it's a perfect icosahedron. We know that's genetically determined, because it's inherited. But what we really want to know is how the equation for an icosahedron is written in the DNA. One could imagine how such an equation could be written on the back of a package of Corn Flakes. It would say: "Cut here, bend there, glue here—and you can fold it all up into this icosahedron." Similarly, viral icosahedrons are made of molecules of proteins packing together in a special way. So if we were to unravel all these structures we would find that the "equation" for a viral icosahedron is written in little bits and pieces in the genome, in a little sequence of amino acids here, and another little bit there. But we can't disentangle this a priori unless we understand the Principle of Construction.

In Brenner's view, it was essential to separate the construction issue—the developmental and building issue—from the functional issue. These are clearly interlocked, and he and Crick believed they could separate these issues and define them.

---

[a] Gedanken (German for "thought") refers to a theoretical or "thought" experiment.

In the summer of 1963, Brenner summarized the outcome of his prolonged deliberations with Crick in a detailed memorandum to LMB director Max Perutz.

Dear Max,

These notes record and extend our discussions on the possible expansion of research activities in the Molecular Biology Laboratory.

First, some general remarks. It is now widely realized that nearly all the "classical" problems of molecular biology have either been solved or will be solved in the next decade. The entry of large numbers of American and other biochemists into the field will ensure that all the chemical details of replication and transcription will be elucidated. Because of this, I have long felt that the future of molecular biology lies in the extension of research to other fields of biology, notably development and the nervous system. This is not an original thought because, as you well know, many other molecular biologists are thinking in the same way. The great difficulty about these fields is that the nature of the problem has not yet been clearly defined, and hence the right experimental approach is not known. There is a lot of talk about control mechanisms, and very little more than that.

It seems to me that, both in development and in the nervous system, one of the serious problems is our inability to define unitary steps of any given process. Molecular biology succeeded in its analysis of genetic mechanisms partly because geneticists had generated the idea of one gene-one enzyme, and the apparently complicated expressions of genes in terms of eye color, wing length and so on could be reduced to simple units which were capable of being analyzed. Molecular biology succeeded also because there were simple model systems such as phages, which exhibited all the essential features of higher organisms so far as replication and expression of the genetic material were concerned, and which simplified the experimental work considerably. And, of course, there were the central ideas about DNA and protein structure.

In the study of development and the nervous system, there is nothing approaching these ideas at the present time. It is possible that the repressor/operator theory of Jacob and Monod will be the central clue, but there is not very much to suggest that this is so, at least in its simple form. There may well be insufficient information of the right kind to generate a central idea, and what we may require at the present is experimentation into these problems. ...

Our success with bacteria has suggested to me that we could use the same approach to study the specification and control of more complex processes in cells of higher organisms. As a first stage, I would like to initiate studies into the control of cell division in higher cells, in particular to try to find out what determines meiosis and mitosis. In this work there is a great need to "microbiologize" the material so that one can handle the cells as one handles bacteria and viruses. Hence, like in the case of replication and transcription, one wants a model system. For cell division, in particular meiosis, the ciliates seem the likely candidates. Already, in these cells, the basic plan of meiosis is present and there is no doubt that the controlling elements must be the same in ciliates as they are in the oocytes of mammals.

Another possibility is to study the control of flagellation and ciliation. This again is a differentiation in higher cells and its control must resemble the control in amoebo-flagellates.

As a more long-term possibility, I would like to tame a small metazoan organism to study development directly. My ideas on this are still fluid and I cannot specify this in greater detail at the present time.

As an even more long-term project, I would like to explore the possibilities of studying the development of the nervous system using insects . . .[5]

Brenner's fancy was captured by a litany of exotic organisms. He was struck by a green alga called *Acetabularia*, a gigantic single cell organism whose size lends itself well to isolating subcellular constituents. He began to tinker in the laboratory with *Caulobacter*, a bacterium with an extraordinary life cycle that involves polar growth. One side carries a stalk to which the bacterium attaches. When it divides one of the daughter cells makes a flagellum, a tail, and the other one makes another stalk. What determines this choice, he asked himself. Other organisms briefly intrigued him, each with its own appealing features of specialized development and function that might be experimentally tractable and informative. Unlike the majority of mid-20th century biologists, Brenner was not paralyzed by the diversity of living organisms. As one commentator stated: "He sought the universal beneath the swirl of particulars."[6]

As he surveyed the animal and plant kingdoms, Brenner began to consider a number of critical requirements. Of central importance were ease of growth under laboratory conditions and a reproductive mechanism that would facilitate contemporary genetic manipulations. Brenner also demanded an organism that would lend itself to the generation of conditional mutants, which he had exploited so successfully in bacteria. In a final bold undertaking, he decided to map the developmental origin and ultimate location of each and every cell in an organism. This approach would be especially valuable in the nervous system for constructing a "wiring diagram" or map that revealed all neuronal contacts. Brenner realized that such a feat would require a level of anatomical resolution that could only be satisfied by an instrument as sensitive as the electron microscope. The electron microscope would also facilitate building three-dimensional models of organs and their relationships to each other, similar to studies he had carried out during his student days in South Africa. The resolving power of electron microscopy is enormous, but, because one can examine only a small bit of anything at a time, the instrument has a very tiny window. The model organism of choice would be, therefore, very small.

Most of the organisms that attracted Brenner's attention were ultimately abandoned because of limitations of one sort or another, related either to the intractability (or lack) of their genetics, the difficulty in propagating the organism in the laboratory, or the fact that its sexual cycle was too slow for laboratory work. The majority of candidates was eliminated as they were simply too large to map cell lineages and fates by electron microscopy. Smallness increasingly became Brenner's primary concern.

Ultimately, he alighted on a phylum of earthworms, the *Nemata*, commonly referred to as nematodes or roundworms, of which there are an estimated 100,000 to 10,000,000 species. Collectively, nematodes comprise by far the most numerous multicellular organisms on earth and can reach sizes of 8 meters. But the great majority are considerably smaller, measuring under a centimeter in length and having less than 1000 cells in total. This latter group eventually became the focus of Brenner's attention.

Brenner had scoured the world in search of the ideal nematode. At one time he even considered obtaining worms from somewhere near the equator. "Surely they must be used to growing at 37°C," he reflected. "Then we can get mutants that grow at lower temperatures and we can do the conditional-lethality thing more easily." With his friend Nathaniel Mayer Victor Rothschild (Third Baron Rothschild), then Director of Research for Shell Oil, he examined a map of the world to see where the oil facilities were located. There were, indeed, a number on or near the equator, but like many of the schemes that flood his brain, Brenner did not pursue this one further.

He also collected a large number of worms around Cambridge, and when someone from the laboratory went on holiday, Brenner instructed them to "bring back some soil!" Even his children joined in, "collecting wholly useless earthworms, but by no means discouraged from this activity."[7] Brenner systematically examined about 60 nematodes before deciding to invest his scientific future (and that of many colleagues and collaborators) in a tiny worm, a free-living soil creature called *Caenorhabditis elegans*.

It was common knowledge that nematodes ate other nematodes, bacteria or fungi. Brenner was intrigued to learn that Ellsworth Dougherty, a scientist in the Department of Nutritional Sciences at the University of California at Berkeley, knew how to grow the little creatures in a defined liquid medium. In early October 1963, Brenner wrote to Dougherty.

> Dear Dr. Dougherty,
>
> I am planning to start work on a small metazoan and, from the work you have done with free-living nematodes, it seems that these might be good organisms for me to start with. I am therefore writing to ask you for a culture of Caenorhabditis elegans, Bristol strain. It would be best for you to send a monoxenic culture since we had best try to axenise it ourselves. Reprints of your papers would also be most acceptable, especially those dealing with nutrition and the recent paper in Science 141, 266, 1963.
>
> Since I have never done anything in this field I hope you will not mind if, from time to time, I have to write to you for advice. One point that puzzles me at the moment is how one gets males, and if one has them, is it possible to propagate them by crosses with the hermaphrodite?
>
> Yours sincerely
> Sydney Brenner[8]

In short order Brenner received nematode cultures from Berkeley with detailed instructions for propagating them in a defined medium, or on Petri dishes sown with lawns of bacteria. *C. elegans* has many features that suited Brenner's needs. It can be grown under controlled laboratory conditions and has a short life cycle, progressing from egg to fully grown organism in a mere three days, a feature essential for genetic analyses in a reasonable time-frame. Most adult worms are hermaphroditic and can generate several hundred offspring by self-fertilization. Every so often, however, hermaphrodites yield pure males, thus facilitating conventional genetic analyses. Both larvae and adults of the organism are transparent, so that its 959 cells can be examined by conventional microscopy. Finally, when grown under laboratory conditions, *C. elegans* offers a number of facile phenotypes that are amenable to genetic correlations, including locomotion, feeding, reproducing, and sensing its environment.[9]

In October 1963 Brenner communicated his decision to launch the *C. elegans* project in a proposal to the MRC that occupied but a single sheet of paper.

> Part of the success of molecular genetics was due to the use of extremely simple organisms which could be handled in large numbers: bacteria and

bacterial viruses. The processes of genetic replication and transcription, of genetic recombination and mutagenesis, and the synthesis of enzymes could be studied there in their most elementary form, and, having once been discovered, their applicability to the higher forms of life could be tested afterwards. We should like to attack the problem of cellular development in a similar fashion, choosing the simplest possible differentiated organism and subjecting it to the analytical methods of microbial genetics.

Thus we want a multicellular organism which has a short life cycle, can be easily cultivated, and is small enough to be handled in large numbers, like a microorganism. It should have relatively few cells, so that exhaustive studies of lineage and patterns can be made, and should be amenable to genetic analysis.

We think we have a good candidate in the form of a small nematode worm, Caenorhabditis briggsae, which has the following properties. . . .

To start with we propose to identify every cell in the worm and trace lineages. We shall also investigate the constancy of development and study its control by looking for mutants.[4]

"There are several things about this proposal which seem astonishing with hindsight," science writer Andrew Brown commented.

The obvious oddity is that the strategy set out in the last two sentences, almost as a throwaway, took more than twenty years to carry out. But that is not what makes scientists giggle about it today. They are used to projects which grow unmanageably; and besides, this one turned out all right in the end. No, the one thing that everyone who remembers this [proposal] pointed out was that it was a grant application on one piece of A4 [paper], asking for, and getting, a quite open-ended commitment to fund largely unspecified research. It is the sort of thing that simply could not happen today, but is also an illustration of the style that made Cambridge such an effective center for molecular biology in the Fifties and Sixties.[10]

The MRC's acquiescence to Brenner's intrepid proposal reflects more than the forward-looking attitude of its administrative leadership. That leadership was acutely aware that Brenner was in a transition phase in his scientific career and that he was casting about for new territory, areas that could be explored just as readily in research enterprises outside the LMB. Notable, too, although Crick's departure to the Salk Institute in La Jolla was still about a dozen years away, the switch from studying simple bacteriophage/bacterial systems to more complex organisms marked the formal end of Brenner and Crick's long and historic scientific partnership.

16

The Many Faces of the
C. elegans Project

Loose gangs

A S WORD SPREAD OF BRENNER'S NEWFOUND SCIENTIFIC PASSION, many were amazed, some disparagingly so. As one observer put it: "Here was a man who had been swimming the deepest and least-chartered seas of thought in modern biology, setting off on an apparently endless trudge across a desert of routine."[1] However, there was nothing in his grand plan that was routine to Brenner, except perhaps his determination to show the scientific world that "still using toothpicks, Petri dishes—and a powerful microscope, you can open the door to all of biology."[2]

The *C. elegans* project was launched on multiple fronts. One of Brenner's primary aims was to map the developmental lineage and ultimate anatomical location of each of the 959 cells that comprise an adult worm. The nervous system was slated for special attention. Accordingly, Brenner elected to construct a comprehensive map of all cell–cell contacts in the nervous system—a "wiring diagram" as he referred to it. He was convinced of the need for a complete wiring diagram of the brain to defend criticism from skeptics.

> If you explain to the skeptic: "I have modeled this behaviour and we've got this oscillator which interacts with this and it's coupled in this way and it does this"; and he says: "That's very nice, but how do you know there isn't another wire which goes from this point, you know, goes right around the back and comes in at the side again?" You need to be able to say: "There are no more wires. We know all the wires."

At the same time as the cell map and the wiring diagram of the nervous system were being assembled, Brenner implemented plans to generate

167

mutant worms with defects in nervous system function. These mutants would be identified by some sort of observable behavioral phenotype, such as abnormal locomotion or feeding behavior. Further, Brenner intended to determine which cells in the nervous system were affected and which genes were mutated. This was the new grand scheme.

To initiate this ambitious program, Brenner required the help of individuals with skills in electron microscopy and computer science, skills then not widely available in the United Kingdom—or anywhere else for that matter. Nichol Thomson, a talented electron microscopist, was the first to join Brenner and his trusted technician Muriel Wigby in the *C. elegans* adventure. Thomson came to Brenner's attention through Victor Rothschild, who joined the Department of Zoology at Cambridge after World War II and became an accomplished developmental biologist. He and Brenner developed a friendship that endured to the time of Rothschild's death in 1990.

Prior to ending his scientific career in the mid-1960s, Rothschild had studied spermatogenesis in the sea urchin. These studies involved extensive electron microscopy that was ably executed by Thomson, earning him the distinction of being "one of the first men to master the crafts, perhaps the dark arts, of preparing specimens for the electron microscope, and soon one of the best at getting pictures out of it."[3]

When Rothschild terminated his career in biology to manage the Rothschild financial empire in the United Kingdom, he was anxious to secure a job for Thomson. When Brenner heard of the opportunity, he hastened to bring Thomson's electron microscopic skills to the *C. elegans* project. A dour Scotsman given to few words, Thomson was singularly devoted to the many challenges that electron microscopy posed in the early 1960s. In due course he became expert at cutting serial sections of the tiny worm.

There was no shortage of other challenging tasks that confronted Brenner and his small research team. In addition to the time and effort spent in learning how to grow and mutate the tiny *C. elegans* and in processing it for electron microscopy, Brenner needed a means of storing the large collection of mutant strains he hoped to generate. Of course, this had been an easy operation when he was working with bacteria. One simply froze the strains in glycerol and tucked them away in a refrigerator. Bacteriophage are equally easy to store, preserved in tiny tubes holding just a few drops of a phage suspension. Now a means of storing worms was urgently required. John Sulston, who joined the lab in the late 1960s, eventually worked out a reliable protocol for freezing *C. elegans* in a state that allowed them to be thawed as viable organisms.

In addition to mastering techniques for building wiring diagrams, reconstructing cell–cell contacts, and growing and maintaining stocks of worms, the group developed clever tricks for identifying mutants. Brenner himself collected the first batch of mutants, those that manifested altered behavior following exposure to mutagens. Subsequently, Muriel Wigby took over this work.

C. elegans has a very brief life cycle, but the effort required to examine hundreds of thousands of worms, to identify the small number that might be mutants, and then to confirm those by genetic crosses, hugely exceeded the effort involved in screening phage plaques. But with dogged persistence, the team slowly identified nematodes with reproducible mutant phenotypes. For example, by placing several thousand worms, previously exposed to one or more mutagens, on one side of a Petri dish and bacteria on the other side, Brenner identified some behavioral mutants. Normally, the nematodes smell the bacteria and move toward them in anticipation of a hearty meal, but those with defective movement lag behind and are easily spotted. By the time he documented his collection of worm mutants in the literature in 1974, Brenner had identified over 100 examples that he called *unc*, for *unc*oordinated. Many of these turned out to have muscle defects, raising the possibility of identifying and cloning genes that controlled muscle development and function (see Chapter 19). Brenner also obtained about every drug known to treat nematode infections in animals and used them to screen for drug-resistant variants. Thus was the worm project was launched.

Performing genetic analyses with a hermaphroditic organism like *C. elegans* was infinitely more challenging than working with bacteria and bacteriophage, even for someone with Brenner's legendary genetic skills. There was no genetic linkage map, and even the number of chromosomes in *C. elegans* cells was unknown. In time these issues were resolved, and many of the skeptics, who had earlier concluded that Brenner was off on a crazy goose chase, began to sit up and take notice.

Many now acknowledge that establishing *C. elegans* as a tractable genetic system was one of Brenner's greatest achievements. Indeed, when he ultimately published the results of these studies in a classic work bearing the title The Genetics of *Caenorhabditis elegans*, he drew attention to the complexities associated with the worm's quirky sex life. "This paper reports the characterization of a large number of mutants, mostly affecting behavior," he wrote. "About one hundred genes have been mapped onto six linkage groups. The methods used are given in some detail, mainly because hermaphrodite genetics has special technical problems."[4]

Data storage was another logistical challenge. Brenner realized at the very outset that he would need to record a huge amount of data, a need that demanded a computer. Recall that this was the mid-1960s, when computers were both scarce and expensive by today's standards. Not surprisingly Brenner's request for a computer solely dedicated to his laboratory was looked upon with dismay. In those days computing was largely performed in computer centers. Requesting the purchase of a dedicated machine devoted to a single scientific project was something of an anathema to the MRC administrators. "And of course if one requested an American computer one could absolutely forget it!" In due course Brenner acquired a reasonably advanced English computer called the Modular I.

Housing the computer in the crowded LMB engendered another tussle with the MRC administrators. Brenner identified (and co-opted) a floor of unfinished shell space in the LMB building that was being covetously reserved for future needs. The entire floor had nothing except basic electricity and plumbing. "This was the most wonderful space I have ever worked in. One had to be neurotic about dust that might get into the computer, so we cleaned out half the floor and lined it with paper. We placed the computer in the middle of the floor and enjoyed the acres of space."

As we saw in Chapter 3, Brenner's interest in computers dates from the time of his association with Seymour Papert in South Africa and was reinforced by his reading about the legendary computer pioneers John von Neumann and Alan Turing. He now became a firm proponent, spending hours and hours learning the ins and outs of computer function, even writing operating systems for his machine. Indeed, he once readily admitted that for about 18 months he more or less disappeared into the computer.

Enter John White, the next strategic team member to join Brenner, Muriel Wigby, and Nichol Thomson in the worm project. Like Thomson, White had a less than traditional education. Drawn to all manner of gadgets ("bombs and rockets and radios—that sort of thing")[5] for as long as he could remember, White elected to postpone a university education after completing high school. Instead, he worked as an electronics technician in one of the MRC laboratories. Sharing his time between gainful employment and studying at Brunel University, White obtained a degree in electrical engineering and was in search of a vocation that offered more interesting opportunities than building electronic gadgets for others. But the job market was tight for someone with his niche expertise, and, on the brink of accepting a position involving nothing more interesting than installing computers in telephone exchanges, he learned of Brenner's computer needs.

Years later White described his first meeting with Brenner in graphic detail. Having acquired a brand new suit for the occasion, he was taken aback to be "confronted by a strange-looking, chain-smoking individual wearing jeans and a striped tee-shirt."[5] Brenner, of course, regaled the bewildered White with his vision for solving the major problems of developmental and neurobiology using a small worm, leaving the young engineer's head reeling as he made his way back to London by car. As he drove he recollected "fragments of Sydney's stream of consciousness"[5] in his mind and an appealing logic emerged:

> [S]tudy the nervous system in a small animal with a few hundred nerve cells, so that the whole neural circuitry could be defined from reconstructions of electron micrographs, deduce the cell lineage of the whole animal by following the development of live embryos using a newly-developed optical sectioning technique, and use genetics to reveal biochemical pathways.[5]

By the time he arrived home, John White had decided that he would join Brenner's laboratory, despite his conviction that the man was "a little mad."[5] Before he could convey this decision, he received a phone call from Brenner. As White was already an MRC employee, it would be a simple administrative matter to transfer him to the LMB. Brenner suggested that White begin work in a few weeks.

Armed with his electronic expertise and familiarity with gadgets, White led the reconstruction of the nervous system of *C. elegans*. He designed a clever, automated worm-slice presenter that aligned serial sections generated by Thomson (a task likened to slicing a worm into salami one-twentieth of a micron thick)[6] for photography under the microscope. He then wrote a program to instruct the computer to collect and reassemble slices. But despite Thomson's best efforts, sections were sometimes unavoidably distorted, generating visual artifacts that were beyond the capacity of the Modular I to correct. Ultimately, White constructed a digitizing tablet and wrote software to drive it so that it was possible to trace sections with the computer and project pictures onto a wall. "Fantastic stuff for 1974! But short of what Brenner and his colleagues hoped for."[7]

In the final analysis, the goal of automatically photographing serial sections of the worm, scanning the photographs, and ultimately generating a three-dimensional model of the entire worm, failed. Regardless, the experience left Brenner with a firm grasp of the enormous value of computers and computation in understanding complex biological systems. "I think that laid the ground for much of the way that I view complex biological systems," he commented. "And increasingly I believe that everyone will have to view such systems that way. Which is, of course, a return to the old Von Neumann paradigm—that if you can't compute something you can't understand it." (Brenner once lunched with the president of the British Computer Society, who expressed his delight in meeting a celebrated computer-savvy biologist. He was interested in examining computer programs written for complex biological systems, in the hope that they might be adaptable for banking. To which Brenner replied, "I was hoping that by studying computer systems developed for complex systems like banking, we might be able to learn something about biology.")

The trio of collaborators Brenner assembled to launch the worm project was as unlikely a group as anyone would expect to find in the halls of the LMB. But Brenner handpicked them with a master plan in mind. Furthermore, they all worked extremely hard—including Brenner himself, who routinely put in 16-hour days that began in the early hours and sometimes went through the night. Nor was that especially unusual at the LMB, where "everything was done in cramped labs, with equipment spilling out into the corridors and the place full of post-docs who never seemed to sleep."[8]

In time the Brenner laboratory began to swell with scientists interested in the many challenges and opportunities that *C. elegans* presented. "People asked what the qualifications were for joining the lab. I told them: 'just an interest in the subject.' " This flexible style of operation attracted people from all kinds of backgrounds, a style to which Brenner attributes much of the success of the LMB. John White, the engineer, was joined by Graeme Mitchison, an algebraic topologist (whose limited contact with biology was through his immunologist uncle Avrion Mitchison), and by John Sulston, an organic chemist by training and another entrepreneurial gadget tinkerer. "I've always found that the best people to push science forward are those who come from outside it," Brenner stated. "Maybe that's the same in culture as well. The émigrés are always the best people to make new discoveries! So when someone once asked me: 'What is the nature of the organization in your laboratory?' I could only think of one answer, which was: 'Loose gangs!' " As the worm project grew it became populated with more traditional inhabitants: graduate students and postdoctoral fellows, all primarily wooed by the infectious enthusiasm that Brenner exuded.

17

Progressing on Multiple Fronts

One felt that the ship was at sea

BRENNER ENTERED THE FIELD OF MOLECULAR BIOLOGY when the experimental models of bacteria and bacteriophage offered the luxury of formulating intellectually challenging questions one day, performing experiments the next day, and addressing new questions the day after that. However, the *C. elegans* project was much more strategic. It took considerable time to mature and could not have been carried out in an environment that demanded short-term gains. Fortunately, the MRC was sufficiently impressed by Brenner's earlier achievements that it continued to support his research, essentially with no interference and with minimal onerous administrative requirements. By late 1968, work was in full swing. A letter to a casual friend provides a summary of where things stood then.

> I started playing with the nematodes in 1964 and have been working inten-
> sively on them for the past three years. My idea was to use them for the study
> of problems of the nervous system, its structure, development and genetic
> specification. I have, therefore, done a lot of things with them, none really
> complete, but enough to tell me that they are going to be very useful indeed.
> Here is a list of some of the things we are doing.
>
> 1. Detailed anatomy by serial section electron micrographs.
>
> 2. Genetics: very advanced. Over 200 mutants have been made and about
>   2/3 of them have been mapped.
>
> 3. Biochemistry: Just beginning. . . .
>
> 4. Drugs: We are screening a lot of compounds and in fact, I am just
>   collecting a set of phosphate ester insecticides.[1]

The first manuscripts to emerge from the *C. elegans* work were submitted for publication in December 1973 and published sequentially in early 1974,

173

a good decade after the project was initiated. The Genetics of *Caenorhabditis elegans* was first in this sequence, with Brenner as sole author. In his introductory paragraph Brenner wrote:

> We know very little about the molecular mechanisms used to switch genes on and off in eukaryotes. We know nothing about the logic with which sets of genes might be connected to control the development of the assemblages of different cells that we find in multicellular organisms.[2]

The labor of establishing the wiring diagram of the nervous system fell to John White and Nichol Thomson, indispensably assisted by the technician Eileen Southgate. Once again Thomson was responsible for cutting serial sections of single worms. Each section was photographed under the electron microscope; after which Southgate traced them to highlight points of interest, making sure that every neuronal connection was numbered and identified. It was not until 1986 that Brenner, Thomson, Southgate, and White published the results of this study in a classic 340-page monograph entitled *The Structure of the Nervous System of the Nematode* Caenorhabditis elegans: *The Mind of the Worm*, a contribution that annotated all 302 nerve cells and the 8000 connections among them.[3] Brenner was disappointed that the final outcome of this gargantuan task was not more illuminating about how genes control the nervous system and behavior. "The right tools for this sort of work are not yet available," he lamented in later years. "That's why I like to have a lot of things going at the same time. So that if one gets stuck with one problem one can carry on with others."[4]

John Sulston joined the *C. elegans* adventure in the late 1960s. "If Sydney Brenner in full flow reminds one of an alpine river, bounding along in an invigorating froth of brilliance, John Sulston has the self-sufficient buoyancy of a dry fly, able to ride the craziest rapids," Andrew Brown wrote.[5] First impressions of Sulston suggest an unassuming individual, decidedly unlike the stereotype of a Nobel Laureate, a distinction that Sulston bears as something of a cross.[a] After completing a doctoral degree in chemistry at Cambridge under Colin Rees's mentorship, Sulston joined Leslie Orgel's laboratory as a postdoctoral fellow in the department of chemistry. In 1964, when Orgel moved to California to direct research in chemical evolution at the Salk Institute in La Jolla, Sulston joined him. He and his wife, Daphne, made an

---

[a] The son of an Anglican priest, Sulston was raised as a Christian, but professes to have lost his faith as an adolescent—a source of considerable distress to his father. See http://nobelprize.org/nobel_prizes/medicine/laureates/2002/sulston-autobio.html.

easy transition to life in California, happily adapting to the laidback culture of the 1960s.

Sulston thrived on tinkering in the laboratory, giving little thought to ascending through the traditional scientific hierarchy or to where he might next land professionally. In the midst of a conversation with Francis Crick (then a nonresident fellow at the Salk), he was somewhat taken aback to realize that Crick was surreptitiously vetting him on Brenner's behalf, with a view to recruiting him to the worm project at the LMB.

"I had heard of this guy at Cambridge who was working with a worm called *C. elegans*," Sulston related. "Everyone was then saying: 'What a laugh! What on earth does he want to work on this little worm for?' "[6] Leslie Orgel, also decidedly skeptical about the worm project, invited Sulston to remain in La Jolla, hinting at the possibility of a more permanent position at the Salk Institute. Sulston and his wife elected to return to the United Kingdom following the birth of their first child. In due course Sulston found himself working with Brenner, whom he described as "very loquacious and very definite in his ideas."[6]

The contrast between the structured environments of his former mentors, Colin Reese and Leslie Orgel, and Brenner's hands-off style of sink or swim required significant adjustments by Sulston. He certainly did not identify with most of the postdoctoral fellows at the LMB, especially the ambitious Americans, whom he described as "in the main totally convinced of their own brilliance and the fact that they were going to carve out brilliant careers."[6] Therefore, after arriving at the LMB in 1969, Sulston took his time in exploring the diverse opportunities of the massive worm project.

The Brenner laboratory had by then isolated and genetically mapped a number of mutations that caused defects in the nervous system and had set about selecting, growing, and genetically mapping more. The serial sectioning of the worm for electron microscopy and model building was also forging ahead. Having had some exposure to neurobiology as a graduate student, Sulston was persuaded by Brenner to initiate studies on a neurotransmitter called γ-amino butyric acid (GABA).

While examining the distribution of GABA-containing neurons in the *C. elegans* nervous system, Sulston was struck by the observation that, compared to older animals, newly hatched worms possessed fewer neurons in the ventral cord (the main nerve pathway that runs the length of the worm). Intrigued by this disparity, he decided to carefully track the lineage of individual cells as the developmental program unfolded. Previous efforts by others to define cell lineages in complex organisms had been frustrated by technical problems. In fact, no one had succeeded in tracking cells beyond the first few divisions of a fertilized worm egg, let alone larvae. However, technical challenges of this sort suited Sulston, and he soon hit upon an ingeniously simple solution. He fashioned tiny pads of agar gel, sufficiently thin so that he could illuminate and focus on individual cells through the gel. Then, placing a worm larva on an agar pad, he covered it with a glass cover slip coated on the underside with a layer of bacteria—worm food! To his delight the larvae remained quite still, feeding on the bacteria, while Sulston was able to watch individual cells dividing. These innovative studies on the lineage of cells in the ventral cord coalesced beautifully with John White's analysis of the anatomy of the ventral cord, providing a fine example of Brenner's grand plan.

Now suitably self-motivated and eager to broaden his horizons, Sulston considered a grander scheme—mapping the entire cell lineage of a *C. elegans* larva. While in the midst of pursuing this formidable goal, he was joined by a new postdoctoral fellow from Harvard, Robert Horvitz. "Bespectacled, intense and extremely thorough in his approach, [Horvitz] arrived at the LMB steeped in the high-tech ambience that he had absorbed as a research student with Jim Watson and then with another pioneering molecular biologist, Walter Gilbert," Sulston wrote.[7]

During his own meanderings through the laboratory soon after his arrival at the LMB, Horvitz noted that Sulston spent many hours peering down a microscope and drawing cells. The Harvard graduate was singularly unimpressed. He wanted to do "real" science, the kind that attacked biological problems at the molecular level and required the use of sophisticated instruments and techniques. Horvitz was thus decidedly skeptical that information gleaned by looking down a microscope could be as reliable and informative as the output from a scintillation counter or from chromatographic and electrophoretic techniques.

Though he majored in mathematics and economics at MIT, Horvitz aspired to a career choice more likely to impact the world than the application of mathematics to economics. In his senior year at MIT, his interest was heightened by formal exposure to biology. Despite his limited knowledge of the discipline, he decided to pursue graduate studies in molecular biology. Indeed, when he entered the graduate program in biology, his laboratory skills were so limited that he asked someone to show him how to use a pipette. In fact, he spent his first couple of months in the lab doing little more than preparing solutions.[8]

Horvitz was particularly stimulated by a graduate course in neurobiology. When one of his MIT colleagues, recently returned from the LMB, told of Brenner and his innovative foray into complex organisms, Horvitz wrote to Brenner to express his interest in exploring learning and memory. Brenner bluntly replied: "Horvitz, I have no reason to think that worms remember or learn anything! How do you feel about suppressor tRNA's?"[8] The two corresponded back and forth, Horvitz persistently pressing his interest in neurobiology. He ultimately received a brief and to-the-point note from Brenner stating: "Horvitz, I have heard from Jim [Watson]; not yet from Wally [Gilbert]. But come and join us."[8] Horvitz arrived at the LMB in early November 1974.

Having established that some *unc* mutants were defective in muscle function, Brenner challenged Horvitz to explore the molecular biology of muscle development. Not especially interested in this line of research, Horvitz began systematically to explore what else was going on in the lab. He soon encountered Sulston, then in the early stages of his ambitious larval cell lineage project. Horvitz was baffled by, if not overtly dismissive of, Sulston's love affair with the microscope. "Where exactly is the experiment?" he rhetorically asked.[8] In time, however, he realized that observations on the cell lineage of the worm might raise interesting and experimentally testable hypotheses. Just as Sulston had observed the loss of cells in the ventral cord, Horvitz's interest was seriously awakened when he noted that some cells died during development. "Here was life and death at the resolution of single cells—and that was exciting to me," he stated.[8]

Sulston and Horvitz teamed up, and by 1976 they had mapped the entire lineage of the *C. elegans* larval state. "Many people wouldn't have seen the point; many still don't," Sulston noted. The two published their results on the post-embryonic lineage of *C. elegans* in 1977 in a classic contribution to the literature entitled Post-embryonic Cell Lineages of the Nematode, *Caenorhabditis elegans*.[9] The treatise consumed 46 pages of the journal *Developmental Biology*. True to his habit of withholding his name from the documentation of studies to which he believed he had made minimal, if any, experimental or intellectual contribution, Brenner did not claim coauthorship.

With postdoctoral fellow Judith Kimble, who was introduced to *C. elegans* by one of Brenner's former trainees, David Hirsh, Sulston and Horvitz shifted their focus from asking *what* happens in the developing worm, to asking *how* it happens. On returning to MIT, Hovitz launched his own laboratory devoted to the study of worm biology. There he made seminal contributions to understanding the phenomenon of apoptosis (programmed cell death), work that won him a share (with Sulston and Brenner) of the 2002 Nobel Prize in Medicine or Physiology.

Like many who worked alongside Brenner at the LMB, Horvitz expresses his unreserved admiration, noting that Brenner

> was as stimulating as anyone could imagine. The coffee room was basically home away from home for Sydney. In the mornings people gathered in the coffee room; at lunch people gathered in the coffee room; and often in mid-afternoon people gathered in the coffee room. And whenever there were people in the coffee room Sydney would appear. He talked about anything and everything under the sun, his brain spinning out idea after idea. Sometimes it was about things that you'd heard once, twice or even ten times. But often it was about something brand new.[8]

As a witness to the historic dynamic between Brenner and Crick, Horvitz was intrigued by the contrasting yet complementary styles of the two heroes of molecular biology. Horvitz stated:

> Sydney and Francis shared an intellectual prowess that was beyond any measurable level, leaving one to conclude that they alone shared some sort of mental stratosphere. They were smart beyond belief; knowledgeable beyond belief; stimulating beyond description.[8]

Brenner's enthusiasm for his work taxed the credulity of even the most dedicated members of the LMB scientific staff. It was not at all unusual for him to work through the night, leaving one to wonder how he survived with so little sleep. Horvitz commented:

> I would occasionally leave the lab as late as 3:00 am and he'd still be there. And he'd be there when I got back later in the morning. He was positively dangerous at night. I'd be working on the cell lineage stuff staring intently through the microscope and for a break I'd go to the tearoom for a cup of tea. Sydney would hear the spoons rattling and come in to chat—at two o'clock in the morning! The perfect time for him to chat of course because there were no distractions. For me it was also a perfect time to go to bed! But of course one didn't walk away when Sydney wanted to chat![8]

Even the week-long festivities attending the Nobel Prize awards could not compete with Brenner's single-minded devotion to science. When Horvitz asked him somewhat incredulously why he was leaving the unique and fun-filled week early, Brenner's response was: "Bob, I'm 75 years old. I don't have time for things like this."[8]

Although a loyal and admiring colleague and friend to this day, Horvitz does not dismiss what has been described by some as Brenner's "darker side." Observing that both men could be impatient, Horvitz judges Crick's toughness to be restricted to those who could stand up to his criticisms. Brenner, he feels, could be more broadly harsh, and if one was of a delicate disposition, one might be seriously upset by things that Sydney sometimes said or did.[8]

Others have commented that working alongside Sydney Brenner can be an intense and sometimes distressing experience. One might work side by side with him for 18 hours a day for a period, only to suddenly discover that he had lost interest in this particular project and was off in a new direction—sometimes never to return. Brenner has often stated that he is not much for the middle game or the end game; he loves the beginning game. A scientific relationship with Brenner can thus quickly vacillate, sometimes leaving a collaborator (especially one much more junior than he) with a disconcerting uncertainty as to whether one is in or out of favor. Such shifts in attitude typically have little to do with Brenner's regard for a colleague as a scientist or even as an individual. The reality is that Sydney Brenner may have some incredibly exciting (to him) idea early in the morning, motivating an inspired but inexperienced postdoctoral fellow or student to begin testing the idea experimentally. If not sufficiently self-confident and knowledgeable, that student may spend the next several years working on some fanciful notion that Brenner once talked about for 15 minutes—and never revisited.

In his account of his own explorations with *C. elegans* Sulston wrote:

> [Sydney's] notion of supervision was to throw out the odd idea and then go away and leave you to it. There would be no weekly lab sessions to check on progress; but he might turn up and quiz you at odd hours. His is a complex and powerful personality ... But he single-handedly started *C. elegans* research and we all learned a lot from him. To be in Sydney's division at the LMB in the 1970s was to be in at the birth of an international community of worm biologists. The great majority of today's worm researchers either came to work with Sydney or are scientific descendants of those who did.[10]

Horvitz and Sulston were among the early worm researchers. However, many more followed, and the worm community now comprises thousands of scientists spread around the world of biology.

By the early to mid 1970s, the nematode program was well established. The basic genetics was completed. Others had begun to generate new mutants and analyze them, and yet others were seeking new avenues to pursue. "One felt that the ship was at sea. But it was very unclear exactly where it was going. The core of it—and it worried me all the time, was how on earth one would get down to finding the molecules involved in regulation. If there were proteins like Jacob-Monod that recognized DNA [referring colloquially to regulatory proteins that turn genes on and off] and if any of our mutants were defective in these, how would we find them?"

18

Getting Back to DNA

*It bridges the biology we practice today and the biology
initiated some fifty years ago*

IN THE EARLY 1970S BRENNER WAS APPROACHED by the Stanford biochemist (and
future Nobel Laureate) Paul Berg. Among other scientific issues, Berg sought
Brenner's advice on the ethics of a series of experiments he was pondering.
Berg was contemplating one of the first experiments in what later became
known as genetic engineering, and certainly did not need scientific advice.
"To my surprise he wanted to know whether or not he should actually *do*
[author's italics] the experiments; whether it was ethically the right thing to
do or not."

Thus, began the recombinant DNA controversy. The contentious debate
surrounding this issue has been the subject of many publications. Two books
are particularly recommended for readers interested in learning more about
that historic period in biology. One, by Donald S. Fredrickson (a central
character in the political storm), is *The Recombinant DNA Controversy, A
Memoir: Science, Politics, and the Public Interest 1974–1981*; the other is *The
DNA Story: A Documentary History of Gene Cloning*, by James D. Watson
and John Tooze.

The experiment that Berg was contemplating was straightforward
enough. He wanted to engineer—more precisely, recombine—SV40 viral
vectors with bits of DNA from another organism and grow these recombinant vectors in the bacterium *Escherichia coli*. The central idea was to determine the feasibility of expressing foreign genes in cellular environments
in which they would not normally be expressed. If successful, the capacity
to move genes around at will offered the potential of a major revolution
in biology, including the exciting possibility of isolating and propagating
(cloning) individual genes for detailed study.

Generating recombinant DNA molecules was technically challenging at that time, but ultimately quite feasible. There were seemingly legitimate concerns about potential safety hazards of the method. In fact, SV40 was known to cause tumors in mice, and some strains of *E. coli* inhabit the human intestinal tract. It seemed entirely possible, if cells carrying the vectors were not properly contained, that they could come to infect laboratory workers, posing the threat of cancer. These initial concerns led to other fears, so that the general public, as well as some scientists, became increasingly alarmed by disturbing scenarios in which cells harboring hazardous genes ran amuck, creating threatening biological aberrations. Fortunately, for all concerned—especially the scientific community—the ensuing years of debate eventually led to a sane (and safe) resolution of the controversy.

Having superficially examined the ethical implications of the studies that he and others were planning, Berg was frankly surprised—even somewhat vexed—when colleagues began cautioning him about the potential dangers of his experiments. However, when his internationally respected Stanford colleague, the late Joshua Lederberg, challenged Berg's dogmatic view about the safety of recombining different genomes in a viral vector, Berg reconsidered his position. Lederberg asserted that these vectors might be readily transmitted to other cells growing in the laboratory, possibly even to humans and other mammals. In light of these warnings, Berg reluctantly decided to cease his recombinant DNA experiments and rethink the issue.

Aside from using viral DNA for genetic engineering, the broader issue arose as to whether health risks were, indeed, associated with working with any tumor virus under standard laboratory conditions. This concern became increasingly pressing as more investigators began to work with tumor viruses in various contexts unrelated to genetic engineering. It was well known that SV40 virus had contaminated the polio vaccine in the United States without evidence of an increased tumor incidence in inoculated children.[a] However, these data were not considered reliable. Record keeping in the United States had been inadequate, and no one could be certain about the cancer risks associated with handling known oncogenic (tumor-promoting) viruses.

---

[a] At various phases of the growth of poliovirus for vaccine production, the infectious agent was passaged through cells (from monkeys) known to be often infected with Simian virus 40(SV40).

In mid-1971 Berg received a telephone call from Robert Pollack,[b] a microbiologist at the Cold Spring Harbor Laboratory, who emphatically sided with Lederberg's cautionary stance. After some discussion Pollock and Berg decided to convene a comprehensive meeting on biological hazards at the Cold Spring Harbor Laboratories in 1973. The meeting produced a small and little-known book on biological hazards,[c] and a series of rather innocuous recommendations for addressing them. Among these was a proposal that blood samples be taken every six months from anyone working with potentially biohazardous material. None of the national hysteria about recombinant DNA had yet surfaced, but it was soon to emerge.

Meanwhile, as recognition of the enormous research potential of recombinant DNA research grew, an increasing number of scientists began exploring the new technology, some surreptitiously. At a Gordon Research Conference in the summer of 1973, Stanley Cohen at Stanford University and his collaborator Herb Boyer at the University of California, San Francisco, described their early work using recombinant DNA technology. Their reports generated considerable excitement about the enormous potential of this technology. However, not all attendees at the meeting were free of concern. Following a heated debate, the cochairs of the conference, Maxine Singer and Dieter Söll, were persuaded to forward a letter to the scientific journal *Science* and to Philip Handler, president of the U.S. National Academy of Sciences. These communications expressed enormous enthusiasm for the new technology, but also voiced significant concerns about its safety.

The National Academy had a long history of monitoring scientific matters in the United States, and Berg soon received a concerned phone call from Handler, seeking advice as to how the organization should respond to this initiative. Berg suggested that a meeting be convened at the earliest opportunity.[1] A select number of relevant scientists were to be included, in particular those with a balanced sense of the scientific implications of the new technology as well as its possible inherent risks. The noted virologist David Baltimore offered to host the meeting of just seven individuals at MIT in April 1974.

---

[b] Robert Pollack is professor of biological sciences at Columbia University and director of the Center for the Study of Science and Religion.

[c] Hellman A, Oxman MN, Pollack R (eds.). 1973. *Biohazards in Biological Research.* Cold Spring Harbor Laboratory Press, Cold Spring Harbor, NY, not to be confused with *Biohazard*, by Michael Rogers (see footnote d).

Following hours of discussion and debate, the committee agreed to forward a letter of measured caution to Handler. Determined to make the most relevant content of this letter available to the public, the group arranged for its publication in prominent scientific journals. An expanded committee comprising the seven attendees at the MIT meeting with three additional prominent scientists composed a letter that was published in *Science*, *Nature*, and the *Proceedings of the National Academy of Sciences*. The letter took a decidedly definitive position, calling for a worldwide moratorium on recombinant DNA work until its possible risks were thoroughly examined. This historic document was henceforth (and often notoriously) known as "the moratorium letter."[1]

Commenting on this initiative, the English scientific weekly *Nature* published an editorial on July 19, 1974:

> In an unprecedented move, the United States National Academy of Sciences has called for a moratorium to be placed on an area of scientific research because of potential and unpredictable hazards to human health. A statement drawn up by a committee of eminent biomedical scientists and released by the academy this week calls for a temporary halt on two types of genetic engineering research because of the risk of infecting man with bacteria containing hybrid DNA molecules whose biological properties cannot be predicted in advance.[2]

The moratorium letter also called for an international conference to address genetic engineering in detail. However, regulatory authorities in the United Kingdom sought more immediate action. To the pleasure of some, but the serious consternation of others, the Medical Research Council (MRC) ordered an immediate freeze on all recombinant DNA experiments in Britain. The Council also assembled a commission headed by Lord Eric Ashby, a distinguished British plant physiologist and Master of Clare College at Cambridge University. Highly regarded for his even-handed style in addressing scientific issues that affect society, Ashby was expected to bring this talent to bear on the deliberations concerning recombinant DNA.

Sometime in 1974 Brenner received a letter from the Secretary of the MRC enjoining him and others to heed the moratorium. Brenner was dismayed. Firm in his conviction that the new technology would offer enormous opportunities to tackle the genetics of higher organisms, Brenner was unequivocally for it. However, he did not realize then how problematic gaining official permission to proceed with this work from the appropriate regulatory authorities would be. He was also unaware how much of his time would be consumed, especially when the Ashby committee was constituted in October 1974.

The Ashby Working Party was charged with the tasks of assessing the potential benefits and practical hazards of techniques that allow the genetic manipulation of microorganisms and of reporting their findings. Brenner was not a formal member of the Working Party. Nevertheless, very soon after it convened, he was invited to appear before the group as an expert witness. In anticipation of this event, Brenner invested considerable time and energy in preparing a detailed white paper.

> The recent development of biochemical methods for joining DNA molecules together with their subsequent introduction into bacteria or animal cells has produced a qualitative change in the field. It cannot be argued that this is simply another, perhaps easier, way to do what we have been doing for a long time with less direct methods. For the first time, there is now available a method which allows us to cross very large evolutionary barriers and to move genes between organisms which have never had genetic contact. ...
>
> When we come to consider the hazards of this kind of work we enter into a very difficult area. ... It is therefore necessary to consider in detail how this work might be controlled. An outright ban would be a disaster, since it would deny us access to the very scientific knowledge we urgently need. Many microbiologists have pointed out that there are containment procedures which can be applied to lessen the dangers. In addition, I believe that much can be done and should be done by molecular biologists to engineer the microorganisms used so that their chances of survival outside a laboratory environment are small. By multiplying a set of small probabilities we could reduce the hazard to very small numbers, say 10^{-24} (reciprocal of Avogadro's number) or even 10^{-50} (about the reciprocal of the number of elementary particles in the universe).[3]

Concerned that control of the volatile issue of genetic engineering might be usurped by the British, Berg rushed to assemble an organizing committee to plan an international meeting in Asilomar, a secluded setting on the shore of Monterey Bay in Pacific Grove, California, and a well-known venue for scientific meetings. Having read a preview of the Ashby report, Berg was impressed with Brenner's style and influence and invited the British molecular biologist to join him on the organizing committee with David Baltimore, Maxine Singer, and the American microbiologist Norton Zinder. The meeting convened in February 1975, by which time the Ashby Committee had already established formal guidelines for recombinant work in the United Kingdom, which Brenner brought to the Asilomar meeting. His intention was to warn the Asilomar attendees of the serious public backlash that might ensue if the meeting projected a laissez faire attitude. Such an attitude would be perceived as provocative, especially by the media, which was to be well represented and was perhaps predictably anticipating the prospect of a likely contentious outcome.

The news media that descended on Asilomar included Michael Rogers, a political writer for the magazine *Rolling Stone*. Rogers devoted an entire book to the conference,[d] successfully capturing the tensions surrounding the meeting and the personalities of the many colorful scientific luminaries—including Brenner, of course. He wrote:

> Brenner is a compact Englishman ... with bushy eyebrows, gleaming eyes, and a nonstop animation, all of which blend to form an impression midway between leprechaun and gnome. Brenner, a late addition to the organizing committee, soon emerged as the single most forceful presence at Asilomar. Over the three days [of the conference] that remained, whenever the sessions wandered off into some technical morass that threatened to swamp the large concerns, Brenner deftly put them back on the track. "Does anyone in the audience believe ... that his work can be done with absolutely no hazard?" He waited through the long silence. "This is not a conference to decide what's to be done in America next week," he continued finally. "If anyone thinks so, then this conference has not served its purpose." ... It was the first time Brenner had spoken out at the conference, and the effect was undeniably tonic.[4]

The primary goal of the Berg committee was to achieve a consensual adoption of two key deterrents both to thwart government intervention and to mollify public opinion. One of these deterrents addressed safety facilities that should be advocated—possibly mandated—to guarantee the strict physical containment of recombinant DNA experiments in the laboratory. The other addressed concerns about how to cope with the inadvertent release of engineered organisms from the laboratory. The latter issue was laid to rest by Brenner's ingenious suggestions for achieving what he referred to as biological containment. Notably, experiments judged to present any risk of human infection with bacteria into which recombinant DNA molecules had been introduced would require the use of strains that were genetically modified or disabled, to preclude their ability to survive outside the laboratory. Brenner's forceful personality and skill in microbial genetics helped convince

---

[d] Rogers M. 1977. *Biohazard: The struggle to control recombinant DNA experiments, the most promising (and most threatening) scientific research ever undertaken.* AA Knopf, New York.

the assembled scientists that safe plasmids, phages, and cells could also be developed.[5]

He was especially helpful when the organizers assembled late on the final night of the meeting to write a report to present to the conferees the following morning. This task did not reach a satisfactory conclusion until the sun was about to rise. "When we were eventually done we were totally exhausted," Berg wrote in a tribute to Brenner. "But we were also so euphoric about having orchestrated a meaningful and constructive report that we took a long walk on the beach in the fading moonlight. We felt that we'd just written the Declaration of Independence!"[1]

David Baltimore is equally effusive about Brenner's contributions. Like the great majority of the Asilomar conferees, Baltimore did not really believe that science and society were threatened by the horrors of recombinant DNA experiments gone awry. He also realized the folly of expecting the rest of the world simply to accept that viewpoint. "Paul, Maxine [Singer] and I realized that we absolutely had to go through this formal process—and Sydney was a real trooper in that sense. Of course his sense of humor was always evident. He once suggested that we consider combining the genes of an orange and a duck—to make duck à l'orange!"[6]

There were occasions during the conference when the sentiment for doing nothing that might impede research was in sharp conflict with the realization that practical solutions for dealing with safety concerns were essential. On these occasions the conferees had to be reminded of the consequences of doing nothing, notably, public condemnation of seemingly self-serving behavior with consequent government interference and crippling legislative prohibition. At one of these turning points, Brenner admonished the die-hards to reject the pretence that there was no biohazard problem and appealed to them to resist the self-serving temptation to reach compromise solutions that would selfishly spare their own areas of research. "What do you suppose will happen if we leave here without doing anything, thereby sending a message to the world that we don't really care?" Brenner iterated and reiterated.

Brenner privately shared the view that the public uproar over recombinant DNA technology was blatantly ridiculous, but he never lost sight of the enormous power of the media in the United States, especially the press. At the very outset it was agreed by the assembled conferees that speakers could request suspension of any recording of the proceedings while they proffered remarks. However, when the audience was asked whether turning off recording devices should include the press, Brenner was the sole

vote in favor. At a press conference later that day, he was confronted
by an indignant reporter from the *Washington Post*, who challenged
the effrontery of a foreigner in questioning freedom of the press in the
United States.

> I had of course appreciated when Richard Nixon left office that the
> press had actually succeeded in getting rid of a President of the United
> States. But I hadn't fully appreciated how arrogant organizations like the
> *Washington Post* had become and how unbelievable their dreams of
> omnipotence were. So I told the reporter: "I believe in the inalienable
> right of adult scientists to make fools of themselves in private and I voted
> this way because I was dead against the press being here." The man went
> berserk—accusing me of being a fascist and so on.

Brenner's perspective was that the moratorium should never have been
called in the first place because it prompted at least one country (the United
Kingdom) to ban recombinant DNA research before the pros and cons had
been seriously considered. At Asilomar, therefore, he forcefully urged suspension of the moratorium. "One should never announce a moratorium
unless one clearly defines the conditions under which it will be reversed,"
he argued. Jim Watson was of the same mind set. Indeed, when Brenner
asked Watson why he had signed the moratorium letter, Watson sheepishly
admitted, "I was a jackass to do that." Toward the end of the meeting, Brenner
offered a formal motion that the meeting declare the moratorium at an end.
The motion passed.

Discussion about real (and imagined) risk categories associated with
recombinant DNA work sometimes bordered on the ludicrous. Brenner
scoffed:

> Clearly malaria is a dangerous disease; so people argued that if one
> wanted to clone malaria DNA one should do this under a very high
> safety category, without taking into consideration what it would actually
> take to reconstitute malaria from its component DNA molecules. On
> those grounds one would have to clone lion DNA at a much higher safety
> category than pussycat DNA because clearly lions are more dangerous
> than pussycats!

Brenner was particularly critical of discussions about the need for elaborate and expensive containment facilities graded from P1 to P4,[e] depending
on the specific experiments under consideration. "There was all this talk

---

[e] Facilities designated as P1, P2, P3 etc. ("P" presumably for "P"hysical containment) defined
increasing levels of containment stringency for recombinant DNA work.

19

Gene Cloning and Genomics

I have too much to do and life is too short

T HE TIME FOR THE RECOMBINANT DNA INITIATIVE could not have been more propitious. In 1977 teams led by Fred Sanger, now Brenner's colleague in the LMB, and Walter Gilbert at Harvard independently invented methods for sequencing DNA. In the same year, Phillip Sharp and Richard Roberts separately discovered introns (stretches of DNA that interrupt some eukaryotic genes).

Stalwarts of the new technology knew full well that the realization of cloned genes would require (among other technical developments) the ability to reduce the huge genomes of eukaryotic organisms like *Caenorhabditis elegans* to manageable chunks that could be stored, propagated, and manipulated. Achieving this goal required versatile cloning vectors. By late 1976 Brenner and his colleagues had invoked conventional genetics to demonstrate that a gene called *unc-54* was the likely structural gene for the myosin heavy chain. This gene became the target of their first foray into eukaryotic gene cloning.[1,2]

One of the many challenges was to confirm that any cloned fragment of DNA included the gene of interest. Simple physical verification using nucleic acid probes for hybridization was not yet routine, and investigators had still to rely on clever genetic tricks. In the Brenner laboratory chemical cleavage of the abundantly available myosin heavy chain protein demonstrated an internal deletion in a mutant form of the protein, an observation that facilitated identification of the *unc54* gene. "We knew we had the right gene because we had done biochemistry on the protein and shown that it had a deletion. When we cloned and sequenced the gene we found the corresponding deletion in DNA."

Brenner expanded his cloning efforts to include genes from the exotic bacteria *Desulfovibrio vulgaris*[3,4] and a strain of *Arthrobacter sp.*, again employing genetic tricks to confirm cloning specific genes of interest.[5] The Brenner group of diligent postdoctoral fellows, including Jonathan Karn, an American molecular biologist, used a series of novel phage λ cloning vectors that facilitated the isolation and analysis of large genomic DNA fragments.[2]

The wealth of expertise in gene cloning accumulated in the Brenner laboratory between the late 1970s and mid-1980s led to detailed studies on the genome of *C. elegans*. Indeed, as early as 1974, the second comprehensive publication on the worm, entitled The DNA Content of *Caenorhabditis elegans* in the journal *Genetics* (John Sulston's first quest in genomics) reported the haploid DNA content of the *C. elegans* genome.[a] Using conventional reassociation kinetics, Sulston showed that the worm genome contains about 20 times more DNA than the bacterium *E. coli*, making it the smallest animal genome then characterized.[6]

Sulston's interest in genomics did not gain serious traction until years later when techniques for characterizing genomes were considerably more advanced. In the mid-1980s, while attending a meeting on developmental biology, Sulston heard a presentation on a technique for physically mapping small regions of a chromosome. The technique, called "chromosome walking," essentially involved "walking" down a chromosome one DNA clone at a time, from a defined starting position. Sulston was struck by the notion that, instead of moving serially down a chromosome one DNA clone at a time, it should be possible to order multiple overlapping DNA clones concurrently, essentially mapping an entire genome in one fell swoop. This strategy offered the added benefit of providing an ordered set of genomic clones for use by the rapidly growing community of *C. elegans* biologists.

Other than his brief experience with measuring the DNA content of the *C. elegans* genome, Sulston had spent the majority of his time at the LMB working on the cell lineage problem. Sulston was encouraged by the possibility of physically mapping a eukaryotic genome for the first time, but was uncertain how to begin. He, therefore, approached Brenner and Jonathan Karn for help and advice.

Karn's primary interest was to exploit genomic libraries for cloning individual genes of interest, in his case the *C. elegans* myosin heavy chain gene. From his perspective the most cogent reason for generating a physical map of the genome was to speed up the cloning work, and he was pleased to

---

[a] Sulston JE, Brenner S. 1974. The DNA of *Caenorhabditis elegans*. *Genetics* 77: 95–104.

join forces with Sulston. By mid-1980 Brenner, Karn, Sulston, and Alan Coulson[b] had generated a set of genomic clones covering the entire C. elegans genome.[7] By mid-1986 the quartet had devised a DNA fingerprinting technique for digitally characterizing fragments of C. elegans DNA 35–350 kb in size that promised to yield a physical map of the genome. By 1989 the map was completed and triumphantly displayed by Sulston and Coulson at the annual "worm meeting" convened at the Cold Spring Harbor Laboratory.

One night at the Cold Spring Harbor meeting, well after the evening session had concluded and the bar in Blackford Hall had closed, Sulston, Bob Horvitz, and two local genomics specialists, Gary Ruvkun and Winship Herr, became deeply immersed in back-of-the-envelope calculations. What would it take to sequence the entire worm genome with existing technology? Just before the sun rose the group concluded that this ambitious goal could, indeed, be realized using cloning techniques that Sulston and Coulson had begun to refine. With Jim Watson's encouragement, they determined that, during the next three years, Sulston would join forces with Bob Waterston at Washington University in St. Louis. The focus of their collaboration was to sequence the first 3 million bases of the 100 million base genome—mainly as a proof of principle. If all went well, they would (with Watson's tacit support) seek funding to complete the genome sequence. Although not directly involved in these plans, Brenner was unreservedly supportive of this grand scheme.[8]

However, somewhere along the line, Sulston's earlier reserved approach to becoming an independent investigator had shifted as his confidence and enthusiasm for a career in science matured. Indeed, in 1989 he had been elected a Fellow of the Royal Society. During his early years at the LMB, John Sulston was perceived as an easygoing ascetic who bicycled to work everyday to reduce pollution. He also protested that garbage collectors should be paid as much as scientists because their work was not as much fun as working in science. "John always gave the superficial impression of being a guru," one of his colleagues commented. "But deep down he is a curious mixture of pure soul, hard drive and dogged determination."[8]

As Sulston's scientific self-confidence grew, he and Brenner began to have different thoughts about issues such as the genesis of ideas and primary ownership of the C. elegans genome work, although these issues were never

---

[b] Alan Coulson had previously worked with Fred Sanger, one of the pioneers of DNA sequencing, thereby gaining valuable experience in molecular biological techniques before joining Brenner's laboratory.

frankly discussed. In failing to appreciate that Brenner's personal interest and involvement in eukaryotic genomics dated from the post-Asilomar period, Sulston misunderstood Brenner's stake in the unfolding genome project. While not directly involved in the *C. elegans* genome sequencing effort, Brenner was delighted that someone from the LMB had the courage to take on this challenge. Furthermore, he certainly considered himself a mentor in the endeavor. As one of their mutual colleagues put it,[8] "John and Sydney settled into different equilibria with respect to their professional relationship." Accordingly, a fallout between the two scientists erupted over an incident that all informed parties agree was rather trivial—one that Sulston himself described as "a mole hill."[9]

Many news articles, principally aimed at the intelligent lay reader, described the exciting discoveries in molecular biology that emerged during the latter half of the 20th century. There was nothing unusual, therefore, in the British science magazine *New Scientist* approaching Brenner for his views on gene mapping. Published in March 1987, the article called Mapping Genes: The Bottom-Up Approach, accompanied by a photograph of Brenner with the caption *Molecular Cartographer Sydney Brenner*, was but a limited section of a broader article about "the race to map the human genome."[c] But Sulston and Alan Coulson were offended that Brenner's casual comments on mapping the genome of *C. elegans* did not include mention of their contributions. The portion of the article in question read:

> Brenner stresses that he wants to make a map, not sequence the entire genome. "I don't think random sequencing of the human genome is on at the moment. We need major improvements in techniques." But technology will improve, he says. "It took three years to map the nematode's genome, but every year we did as much as all the years before. Molecular cartography is going to be a real technology. It will become just a normal scientific endeavour." Already one researcher can map the genome of a bacterium in a few weeks. "Maybe we can get to the stage when you can do a bacterium in a day."[10]

Events took on a more confrontational tone when, a few weeks after the piece surfaced, *New Scientist* published a brief letter authored by Sulston and Coulson (that also carried Brenner's name).

> We would like to clarify two points in your report on mapping the human genome ("Mapping Genes: The Bottom Up Approach" 5 March, p. 36).

---

[c] Joyce C. 1987. The race to map the human genome. *New Scientist* 5: 35–39.

First, the nematode genome is not being mapped in Sydney Brenner's Molecular Genetics Unit, but by Alan Coulson and John Sulston in the Laboratory of Molecular Biology, as would have been made clear had the authorship and journal been cited for the published figure.

Secondly, and more importantly, the map of the nematode has not been completed in three years, nor can a map of a bacterium yet be completed. What has been achieved is the ordering of clones of the genomes into a number of segments (about 1000 for the nematode, 50 for the bacterium). The much more difficult task of joining these together, by finding the rarely occurring clones that link them, will take a great deal longer.[11]

Immediately following this release, Sulston penned a hand-written note to Brenner:

Dear Sydney,

In order to set our relationship on a secure footing for the future, I suggest that we exchange, in confidence, letters of intent. Ours is enclosed. I believe that it merely formalizes normal scientific protocol.

Yours,
John[12]

The specific suggestions that Sulston and Coulson offered in their formal letter of intent to Brenner included the following:

No reference shall be made in the press to the nematode map without our prior agreement, and no verbal statements shall be made about the nematode map without acknowledging our pioneering role.[13]

Brenner was not at all pleased with this unexpected turn of events. A few weeks later he responded to Sulston and Coulson.

. . . I reserve the right to refer to published work without necessarily consulting you first.

Reciprocally, you will acknowledge in your publications the use of my ideas or methods developed by me. Coauthorship should be reserved for projects that are done in collaboration by prior agreement.

I would also like to say that I reserve the right to carry out mapping experiments on the nematode genome with my own techniques. I am willing to inform you of any plans, if you wish, but this should not be construed as seeking your permission in any way.[14]

Over the years Brenner and Sulston have shared the public stage in receiving many scientific awards, including the Nobel Prize. But their relationship has never really healed. Sulston, who professes genuine remorse about the entire incident, asked the rhetorical question: "Would I have done better not to rock the boat?" His answer:

> I don't know. Mostly, I'm painfully aware of how petty the entire thing really was and I wish it had not had the painful outcome that it did. Sadly it's on such tiny molehills that some relationships stumble. I tried to write about this in *The Common Thread* but simply couldn't find the right tone.[9]

Brenner's sentiments about this unfortunate incident are revealing of his attitude to other confrontational situations that over the years threatened to distract him from his intense focus on science. He was upset by Sulston's behavior, but had little interest in dwelling on the issue.

> I really don't understand what John was so upset about. I introduced him to genomes in the first place at a time when he was feeling very insecure about his future at the LMB and was considering leaving. When people behave that way I don't have time to deal with them. I simply move on. I have too much to do—and life is too short to be caught up in this sort of thing.

Ultimately, the newly established Sanger Institute in Cambridge and the Genome Sequencing Center at the Washington University School of Medicine in St. Louis collaborated to sequence the genomes of both *C. elegans* and its relative *C. briggsae*. An essentially complete *C. elegans* sequence was published in December 1998—the first multicellular animal genome to be sequenced. But Brenner's name was not among the dozens of authors. The remaining bits in the *C. elegans* DNA sequence were revealed in October 2002 and a shotgun assembly[d] of the entire *C. briggsae* genome was made available in July 2002.[15]

Sequencing the *C. elegans* genome paved the way for Sulston's involvement in the Human Genome Project, which culminated at the turn of the century with the first draft of the sequence of the three billion base pairs of the human genome. Commenting on the significance of the *C. elegans*

---

[d] Shotgun assembly refers to a strategy in which a genome is broken up to yield random short fragments of DNA. These are then examined for overlaps in nucleotide sequence, allowing the generation of contiguous DNA segments.

sequencing project following the landmark 1998 publication, biologist Martin Chalfie wrote in a *Nature* commentary:

> The benefits of the [*C. elegans*] sequencing project go beyond merely identifying genes and working out the sequence. First, the *C. elegans* groups have shown that complex genomes can be sequenced. . . . Second, the *C. elegans* project has been a model for how an efficient genome effort can be run. . . . Third, the *C. elegans* Genome Project has stimulated the creation of powerful software needed to manipulate the genomic data. . . . Fourth, . . . the *C. elegans* Genome Project is a superb model of how sequencing can best serve the scientific community. When he first described the project, John Sulston remarked that one of his main goals was to promote an open and free exchange of materials. From the start, all data, clones and sequence were freely available.[16]

Sulston became Director of the Wellcome Trust Sanger Institute in 1992, but resigned this position in 2000. He was knighted a year later. He now devotes much of his time to one of his other passions, hiking, sometimes with his wife, Daphne, sometimes alone. Sulston says that he is happiest when walking the English countryside alone, often for days on end, reading and reflecting—and sleeping in his portable tent.[9]

The *C. elegans* project certainly was an ambitious foray into the complex world of multicellular organisms, encouraging a community of *C. elegans* researchers that now numbers in the thousands. But Brenner had hoped for more impressive dividends. In 1973, not too long after the *C. elegans* project was fully underway, Brenner offered a probing analysis of his expectations for the most challenging aspect of the worm project, understanding the nervous system.

> Tracing the connexions between genes and behaviour is a fascinating but very difficult problem. . . . Many of the elementary processes of nerve cell growth are not at all understood at the present moment. Nor do we have much idea of the genetic control mechanisms that may be used in higher organisms. As to how the effects of such genes are mediated is an entirely separate question at the moment. The difficulty is that the implementation problem, as I call it, can be plausibly solved at the molecular level in many different ways. Thus a gene that affects the subset of neurons may specify a protein which is a controlling element enabling that cell to call other genes, which then execute the specific process required for that cell. Or, alternatively, it may determine a protein which is located in the membranes of that subset of cells and enable those cells to respond to an outside signal. . . . The solution of these problems will not be easy in whole organisms and may have to await piecemeal investigation in other systems such as cell culture.[17]

A full decade after this publication Brenner reiterated these formidable challenges in an interview with Roger Lewin.[e]

> We tend to talk loosely about genetic programs and we should be careful about the implications of this language, even when used metaphorically. The total explanation of all organisms resides within them, and you feel there has to be a grammar in it somewhere. Ultimately, the organism must be explicable in terms of its genes, simply because evolution has come about through alterations in DNA. But the representation will not be explicit. We [will] need to understand the grammar of development to make sense of it.[18]

Brenner remains staunchly optimistic that nothing in nature is beyond comprehension by those endowed with sufficient intelligence and tenacity. "Peter Medawar has written that science is the art of the soluble.[f] My stance is that science is the art of the inevitable. If you pursue it with passion and clever thinking it's inevitable that you will find out how everything in nature works. Maybe it won't be until you are able to put the last period on the page; when you know everything, that you will be able to say, 'Aha, now I understand!' But you will get there in the end."

Not long after his interview with Lewin appeared in print, Brenner received a note from James Crow, an early pioneer of genetics.

> I had one horrible thought on reading the article. If, as might happen in the worst of all possible worlds, you learn the complete lineage of every cell, and then the patterns of mutants and all your molecular insights prove fruitless, what do you do then? I can't help thinking of classical comparative vertebrate anatomy and embryology. Perhaps the reasons for some of the illogical developmental patterns will be understood if you know the evolutionary history (just as the seemingly nonsensical development of the mammalian jaw and middle ear make sense when one studies the lower vertebrates). I hate to suggest that the way out is to learn the complete cell lineage of every related invertebrate family; but if so, it will at least provide employment for our grandchildren.[19]

To which Brenner replied:

> I think you are right when you say we will have to know a lot about different animals before we can begin to make sense of development, and then only in the context of evolution. . . . I am always rather pleased when at the end of the day the problem seems more obscure than ever![20]

---

[e] Roger Lewin is a notable American author and science journalist.

[f] *The Art of the Soluble* is a book of essays by Medawar that was published in 1968.

The *C. elegans* project must be viewed as one of the great success stories in biological science. The MRC funded the project in deference to its unwavering confidence in Brenner. Brenner and his initial group of coworkers pressed on for five years without publishing a single paper, and as Andrew Brown pointed out, "for eight years before producing something of which the world would have to take notice. That would be impossible in science today; yet without that freedom the project would never have happened. . . ."[21]

The payoff was satisfying. Brenner and Sulston represented the ninth and tenth Nobels for the LMB in just 34 years, awarded for work carried out there. No other academic or research institute has come close to such stunning recognition.

In a reflective mood in the tranquility of his Ely garden one early autumn evening, Brenner mused about his Nobel Prize. His earlier contributions to prokaryotic molecular biology, especially the discovery of messenger RNA and deciphering aspects of the genetic code, were more intellectually challenging and rewarding than the *C. elegans* work. Brenner sees this earlier work as more worthy of recognition by the Nobel Foundation, and he is not particularly secretive about this view. On more than one occasion, in fact, he has claimed that he is delighted to have been awarded two Nobel Prizes—the first he never received! Brenner provided a revealing response to a congratulatory note from his friend and colleague the late Seymour Benzer. He stated, "[The prize] has to be tied to a discovery, and that was apoptosis. I rode in on the tail of Sulston, who rode in on Horvitz's tail."[22]

The committees that annually anoint a select few with the distinction of Nobel Laureate work in weird and wonderful ways. But it is surely more than just, that Brenner did not fulfill what at one time seemed like a promise of becoming the most accomplished scientist in the world, never to have won a Nobel.

20

Director of the LMB

The biggest mistake of my life

B RENNER ARGUES: "THE MOST WONDERFUL THING IN SCIENCE is the opening game—when there's nothing else there. That, I think, is when one can exercise a tremendous amount of freedom of intellectual choice. And having played the opening game with *C. elegans* it was now in the middle game,[a] with lots of people taking pawns and moving knights around the board. So I thought I'd better find another game!" In the late 1970s Brenner, then at the acme of his scientific career, agreed to serve as Director of the Laboratory of Molecular Biology (the LMB), a decision that he would describe as "the biggest mistake of my life."

Soon after the LMB opened its doors, Francis Crick, then head of the Division of Molecular Genetics, petitioned Perutz to appoint Brenner as joint head of the division so that his administrative load could be shared. It is suggested that in reality Crick artfully divested himself of most, if not all administrative chores, delegating authority and control to his junior partner. Perutz and other senior members of the scientific staff were quick to recognize the merits of Crick's proposed arrangement, prompting Perutz to extend this leadership structure to other divisions. It did not escape Perutz that establishing joint division heads offered a persuasive argument for leadership succession from within a division. This new organizational structure would help to avoid unwelcome scrutiny by the MRC, which had a habit of carefully reviewing units when division heads retired, often shutting them down.

---

[a] Brenner attributes the comparison of scientific research to the game of chess to the famous British physicist Desmond Bernal.

Well known for his grace and charm, Perutz managed the LMB in a low-key manner. Despite conveying an impression of bumbling along, Perutz was known to be appropriately tough and decisive when difficult decisions were required. In fact, Brenner once told an interviewer, "[T]here was Max Perutz the scientist and there was the Archduke Maximilian Ferdinand from old Vienna. . . ."[1] However, managing budgets was not one of Perutz's notable skills. In part this was a consequence of the scientific success of the LMB, prompting a categorically paternal attitude from the MRC head office. Thus for years the coffers were full, and there was little oversight of spending habits, particularly for infrastructure and support facilities such as the lavishly stocked laboratory stores and the LMB library.

The ascension of a conservative government to power in 1970, with Edward Heath as Prime Minister and Margaret Thatcher as Secretary of State for Education and Science, signaled an end to this cozy situation at the LMB. Perutz was scheduled to retire as Chairman of the Governing Board (a title that he preferred to that of Director) in 1974, at the mandatory retirement age of 60. As this date drew near, the MRC announced its intention of conducting a review of the LMB, and to "formulate a Council policy for the future support of molecular biology in universities and research institutes, both in the United Kingdom and abroad."[2]

A 1974 report from a special MRC committee chaired by David Phillips, professor of molecular biology at Oxford University, recommended draconian measures—notably a 25% cut in the budget of the LMB and a thorough revision of its managerial and administrative structure. The "Phillips cuts," as the report was disparagingly known, more or less coincided with the retirement of Harold Himsworth, the popular Secretary of the MRC, and emphatically signaled the end of a postwar bonanza for British science in general. Predictably, the Phillips Report generated much debate and controversy. Members of the LMB obviously favored the relaxed manner in which the lab had been managed since its inception. The Phillips Committee viewed things differently, expressing little confidence "that the existing system, even if it had worked in the past, would work in the future."[3] Many outside the LMB acknowledged that there had long been tensions between various MRC units and the British university system about the perceived favor with which the MRC units were treated. They felt the time had come to correct this state of affairs.

In the interim, Perutz, with the support of his LMB colleagues, successfully petitioned the MRC to extend his directorship for five years, through 1979, a precedent that later was to have repercussions involving Brenner.

This move postponed the vexing problem for the Council of identifying Perutz's successor, but ultimately the decision had to be faced, despite the view in some quarters that he was irreplaceable. One official visitor to the LMB wrote: "I certainly find it difficult to conceive someone coming in with the abilities of Perutz and conducting with his skill the harmonious management of this group of really quite tricky distinguished senior scientists."[4] Intent on a more formal management structure at the LMB, the Phillips Committee recommended that Perutz's administrative structure of a Governing Board led by a chairman be reversed to the usual practice of having a director. There were no obvious candidates for this position, who historically came from within the LMB ranks, and, in the end, Sydney Brenner was tapped for this role.

Brenner was not at all happy with the Phillips report and its implications at a time when the science budget in the United Kingdom was strapped. He especially was not keen on forging formal links with neighboring Cambridge University, as the report recommended. However, he was looking forward to introducing a stronger molecular biological orientation at the LMB. In particular he wished to make the "new biology," developing with the promise of recombinant DNA technology, more visible.[5]

There was brief discussion about the wisdom of letting Brenner assume the directorship in 1977, two years before Perutz was due to formally retire. Nonetheless, Perutz was determined to hand over an administrative structure that was on a more solid footing. Thus, Brenner was appointed proleptic (anticipatory) director in that year, with the objective of succeeding Perutz in 1979, an arrangement that prompted the retort: "During the time I was nominally proleptic director I referred to myself as the epileptic director!" Brenner required some arm-twisting to accept the position because he knew that it would inevitably distract him from his research. "But I accepted the position out of some kind of loyalty to the institution."

In late April 1977, *Nature* published an announcement under the banner, "Brenner's appointment."

> The UK Medical Research Council (MRC) has announced that Dr. Sydney Brenner will become leader of the MRC Laboratory of Molecular Biology (Hills Road) in Cambridge on the retirement of Dr. Max Perutz in 1979. The news is little surprise for those who have heeded the rumours that followed an internal MRC report produced last year on the future of Hills Road, suggesting a radical change in organization. The MRC has followed this, for whereas Dr. Perutz was chairman of the Governing Board of the laboratory, Dr. Brenner is to be its Director. ... In 1979 the Governing Board will

probably be disbanded. Its replacement if any, is currently under discussion, but whatever the outcome Dr. Brenner is certain to end up with a good deal of power.

> All told Sydney Brenner will have a considerable challenge on his hands ... during the course of his Directorship, which automatically lasts ... until 1987, when he is 60 years old. Dr. Brenner was recently heard to say "I am going downhill and I am going with a lot of noise." Only the second part of that statement is true.[6]

Aside from the somewhat reckless spending under Perutz's watch, the budgetary issue was seriously complicated by the reality that the MRC leadership was no longer doling out money simply for the asking. Also, Perutz was predictably viewed as something of a lame duck director for a year or so before he stepped down. Anticipating, therefore, difficult financial times ahead, Michael Fuller, the laboratory manager who enjoyed the quaint title of Laboratory Steward under Perutz, went on a tear, aggressively stocking the laboratory stores in a less than thoughtful manner.

Brenner's appointment was favorably viewed by his LMB colleagues, as they anticipated that he would be generally more careful with the budget than his predecessor had been. Furthermore, as he now enjoyed enormous scientific stature in Britain, it was hoped that Brenner's soaring reputation would help restore some of the financial largesse that the lab had enjoyed for so many years. Finally, there was a strong sentiment—not universally shared—that the time had come for a cell biologist, instead of an X-ray crystallographer, to direct the laboratory and its future.

When Brenner assumed the directorship in 1979, he discovered that the LMB was about a million pounds in the red—not a promising beginning for any new leader. This unfortunate state of affairs was seriously aggravated by a period of double-digit inflation throughout England. "We were seriously under-funded and this [the deficit] had to be dealt with."

Brenner wasted no time in informing the scientific staff of the gloomy financial situation. He invoked a budget freeze and gave notice that for the foreseeable future there would be no spending without his express approval.[7] In Brenner's view, academic administration was largely a matter of common sense.

> For example, I discovered that for the last six weeks of each year essentially nothing was ordered because the laboratory had run out of money. Now of course that was ridiculous, because in six weeks time you were going to have more money. You aren't really in a deficit until you have no money at all and I quickly realized that the best way of operating was in fact to have a

moderate deficit. You certainly didn't want to have any funds left over at the end of the budget year, because these would go back to the MRC. And with the kind of hyperinflation we were dealing with it was better to spend all your money early, because by the end of the year things would cost 25% more!

Brenner officially became director of the LMB on Monday, September 3rd, 1979. But the previous Saturday evening, he suffered a debilitating accident while riding his motorcycle and was forced to assume his new administrative responsibilities from a hospital bed! By no means a motorcycle fanatic, Brenner adopted this mode of transportation to avoid the heavy automobile traffic to and from the laboratory and to curtail the outlay of cash for exorbitantly priced gasoline. While motorcycling in Cambridge on that Saturday, he had spotted a wallet in the street and stopped to retrieve it. The wallet contained a considerable sum of money, and the owner appeared to be German, possibly a tourist. Assuming the owner would like it back, Brenner delivered the wallet to a nearby police station. The afternoon was late and traffic heavier than usual, when a taxi collided with Brenner as he left the police station. The disastrous result was a severe compound fracture of his left leg, involving the lower end of the femur and the top end of both the tibia and fibula. Brenner's knee was essentially destroyed. "It took about fourteen months for this to heal to a point where I could really get around again."

From the outset of his confinement to a hospital bed, Brenner retained his indefatigable energy for work—and his sense of humor. When an attentive surgical registrar (resident) gravely explained the severity of his injury, he plaintively asked the young physician, "Doctor, please tell me, will I ever play the violin?" When the registrar assured him that he most certainly would, Brenner observed, "Oh, that's odd. I never played it before!" Despite his energy, Brenner was not able to resume his hectic travel schedule until well into 1980.

A note from Francis Crick paternally chided Brenner for working too hard and stressed the imperative of a prolonged rest—an atypical attitude for Crick.

> I hope [hospitalization] proves a useful interlude. I have the impression that for the last few years you have been rather too close to what you've been doing, mainly due to the pressure of work, so it wouldn't surprise me if a little forced inaction might restore a broader perspective.[8]

Of course, Brenner disregarded this sane advice. Despite considerable pain and extremely limited mobility, he assumed his directorial and other professional duties within a matter of days, managing both the LMB and his research program from his hospital bed. A colleague commented that, as a result of the accident, he probably knew more about what was going on in the lab than if he were healthy and traveling around the world.[9]

Brenner has been plagued by recurring problems with his leg to this day. In particular, poor healing of recurrent bouts of cellulitis (infection of the skin) have necessitated relatively frequent skin grafting, with further poor healing and an additional cycle of complications. He has had to use a cane ever since the accident, an imposition that has not in the least hampered his enfant terrible persona.

Dissatisfied with the administrative assistant he inherited, Brenner was further frustrated when a replacement that he had identified was not seated. He then turned to the MRC Head Office for help. Fearful perhaps that the LMB might continue its recent downward spiral, MRC Secretary Sir James Gowans (Himsworth's successor since 1977) dispatched Bronwen Loder, an experienced and no-nonsense MRC administrator at the head office for many years. "When Bronwen took over she effectively ran the place," Brenner related. "The MRC wanted stricter oversight, and I suspect they were also pleased to have someone who they knew well in place, to whom they could give direct instruction."[10]

An articulate, administratively experienced and confident woman, Bronwen Loder was not at all upset by her transfer to the LMB. On the contrary, she was pleased to be rid of the machinations that characterized elements of the politically charged MRC head office under Gowans's leadership. When instructed to prepare a job description for the position, Bronwen crafted one that fitted her precisely. She joined the LMB in 1983, understanding full well that "my primary mission was to be the watchdog of the laboratory."[10] An earlier MRC inspection of the LMB, in which she participated, had revealed considerable mismanagement. In particular the inspection team was dismayed to discover that, during Perutz's tenure, the laboratory stores were excessively stocked and poorly inventoried. Boxes and boxes of surgical gloves, all useless because the gloves were powdered on the outside, was one of the more amusing finds. Former LMB director Richard Henderson recalls his surprise when he once visited the laboratory stores to acquire a new pH meter and was casually asked how many he wanted! When he enquired how many the stores carried, his surprise turned to mild shock at the reply: "Oh, about twenty."[9] On another occasion Henderson requested a new anorak for cold room work and was again taken aback to be asked politely what color and size he wanted.[9] Then, too, members of the overstaffed library spent much of their time photocopying articles requested by anyone and everyone.

In accord with her expected duties, Loder demanded strict compliance with MRC rules. Not surprisingly, this action did not meet favorably with some of the senior scientific staff, particularly those accustomed to Perutz's lackadaisical style. On the other hand, she was well versed in the inner workings of the MRC and was able to secure some expensive acquisitions that might not otherwise have been obtained. Indeed, in due course, the MRC began to view Bronwen as someone more supportive to the short-term goals of the LMB than to those of the MRC.[10]

During her two years at the LMB, Loder established a close working relationship with Brenner, who quickly came to rely heavily on her administrative savvy and general toughness. This dynamic did little to endear her to the rest of the scientific staff, some of whom grumbled about a conspiratorial relationship between the two. "I'm sure we hatched the occasional plot—but not many," Bronwen stated.[10] More likely Brenner brought the same proximity and intensity to his new administrative relationships as he did to his scientific relationships at the LMB. In particular, he was used to making decisions on his own and is said by some to have consulted with his colleagues infrequently. Some grumbled that, despite being an effective intellectual leader, Brenner was not overly concerned about or skilled at building morale. As one long-time colleague remarked, Brenner's reaction to someone who did not readily care for the way he did things ranged from being highly perplexed to being really angry. Another commented that Brenner did not like administration when others were administering; when he was in charge, however, he could became quite dictatorial. In short, accustomed as he was to operating according to his own dictates, Brenner was perceived as a maverick—sometimes as a nefarious schemer. This perception was not at all helped by his closed-door relationship with Bronwen Loder. Adding fuel to the flames, some of Brenner's senior colleagues were reluctant to fully embrace Brenner's desire to introduce more molecular biology to the LMB, especially recombinant DNA technology, opinions about which were still controversial.

In the wake of the Phillips report, resources were especially difficult to come by. Past capital improvements aside, space at the LMB was also once again at a premium. Thus, dealing with a group of entrenched staff was challenging, especially for someone with Brenner's unyielding personality. The continued presence in the background of Max Perutz, even now as a scientist without administrative portfolio, did not make life any more comfortable. With regard to Perutz, the MRC rescinded two long-standing rules—that an MRC unit must close when the director retires and that retired directors may not occupy work space in the premises of their former unit.

Brenner's views of his relationship with his colleagues at the LMB during his seven years as director are predictably discordant with those described by others. With the directorship he inherited a federation of three long-established and largely independent research divisions: protein chemistry, protein crystallography (soon changed to structural studies), and molecular genetics—each with its unique style and fiercely guarded tradition, and each managed by senior members of the scientific staff. "My job as the new director was to convert the LMB from a federation to a union," he stated.

> I thought about how to go about this and composed a document that I circulated among my colleagues informing them of what I proposed to do. In it I told people that we had to operate differently with respect to finances. We also had to recognize that there were new and restrictive labor laws in Britain and we had to be more careful about the way we made appointments. I realized too that I had to plan for the long-term future of the lab. Most scientists are only concerned about what will happen next week—or next month. People don't think about next year—or ten years down the road!

Planning for the long-term future was an appropriate and relevant concern. With the approaching decade of the 1980s, the LMB remained a formidable scientific establishment, but the rest of the research world was catching up quickly. New technologies and research trends were emerging in other places, and the steady flow of bright American postdoctoral fellows to the LMB—long a recognized Mecca for training in molecular biology—was slowing.

Embracing the new technologies of gene cloning and protein expression was essential, Brenner thought, for advancing the fields of developmental biology and neurobiology from the merely descriptive to the cutting-edge. Coming as he did from a strong molecular biological tradition, Brenner was convinced these technologies would continue to be fundamental driving forces and was determined to promote their use in the unit. To his disappointment, however, some of his colleagues expressed their frank opposition. Indeed, some signed a letter to Brenner essentially informing him that they did not care much about DNA at all. Nonetheless, Brenner was unswerving in carrying out plans that he believed to be in the best long-range interest of the laboratory. He recruited the forward-thinking cell biologists Kim Nasmyth and John Rogers; at one point the future Nobel Laureates Paul Nurse and Richard Roberts were under serious consideration. A new stream of postdoctoral fellows in nematode biology included future scientific stars such as Cynthia Kenyon, Judith Kimble, Edith Myers, Ed Hedgecock, and Martin Chalfie. But these enlistments did not always sit well with his senior colleagues either.

Brenner gave up his role as director of the division of cell biology and created a new Director's Division. This entity housed his own research space as well as space purposefully kept empty so that he could allocate it at his discretion. Not surprisingly, the sight of unused—but much coveted—space concerned many of Brenner's in-house colleagues.

More serious events were to contribute to Brenner's growing frustration with managing the LMB at the expense of valuable research time. The MRC owned a nearby facility that housed (in Brenner's opinion) a less than distinctive neuropharmacology unit. When the director of that division was due to retire, Brenner made a determined push to establish modern neurobiology at Cambridge. By appointing distinguished new leadership, Brenner hoped in time to incorporate neurobiology seamlessly into the LMB. Thus, a search for a suitable leader was mounted. The possibility of recruiting Max Cowan, an accomplished South African neurobiologist, then provost at Washington University in St. Louis, generated considerable excitement. Despite all efforts, the MRC was unable to match Cowan's American salary—and the initiative collapsed.

When no one else surfaced as an appealing candidate, Brenner changed tactics. Stressing the importance of moving ahead in the neurobiology arena, Brenner proposed to MRC Secretary Gowans a new division of molecular neurobiology in the LMB that would retain selected members of the neuropharmacology unit. To further entice Gowans, he offered to launch the neurobiology initiative by transferring all ongoing *C. elegans* neurobiology efforts to the new division. To tempt Gowans even further, he proposed his intention to include a new program in retinal biology.

In October 1983 Brenner forwarded a pointed memo to the MRC Council addressing this issue:

> The Unit is definitely needed. Research in neuroscience will be a tremendous growth area in this decade, and Britain is not so well supplied with centers for it that we can afford to close or divert one. The spectrum of work at other MRC Units in this general field is not sufficient, in my view, to generate the momentum needed for it in this country. We are poorly placed compared to the USA, and Japan, and even France. .... Having good knowledge of what is going on in our competitor countries in this area I remain convinced that the most profitable line for such a Unit to pursue is molecular neurobiology.[11]

Brenner worked hard to convince his senior colleagues that the LMB would not sustain its impressive international reputation if they failed to chart new research directions—but to little avail.

Some of my colleagues complained that the lab was getting too big and that we shouldn't start new programs. They didn't seem to recognize the importance of neuroscience and the sad lack of its presence at Cambridge University. It was said too that the LMB was getting too big—and that Brenner was getting too big for his boots!

Gowans had apparently supported Brenner's notion of mounting a program in neurobiology in existing LMB space, but then inexplicably (to Brenner) changed his mind. Brenner was infuriated by this perceived broken promise.

> The tactical mistake that I made then was not to suggest to Gowans that I would quit as director of the LMB and take over this unit instead, knowing full well that we would almost certainly do a reverse merger after a while. But I didn't do that, and Gowans, presumably bowing to pressure from others, reneged on his promise to move ahead. I was told that the neuroscience search committee thought I was greedy and were opposed to the move. Gowans voiced the excuse to me that since the committee had opposed this move his hands were tied. When he was initially keen on promoting this idea he didn't consultant any committees. But when he wanted to squelch it he used that well known strategy!

In 1984 a new neurosciences unit under the directorship of Eric Barnard, who studied neurotransmitter receptors, was finally approved by the MRC. Brenner was bitterly disappointed at this outcome and was not hesitant to so inform Gowans in a pointed personal letter.

> Dear Jim,
>
> Thank you for the lunch on Friday [three days before this letter].
>
> I told you then of my disappointment with the decision to give the neurobiology unit to Barnard. I think it is the wrong decision, being neither scientifically credible not particularly practical in that it only delays the final day of reckoning. . . .
>
> Despite your assurances, there was always some doubt in my mind about the outcome. Our bid had stringent (but honest) conditions and the response of the Neurosciences Board and their representatives on Council was predictable. So the failure of our bid did not come as a total surprise to me. What is inexplicable is that you did not tell me sooner that you had embarked on new negotiations with Barnard and that my proposals had already been formally rejected. It will also not be easy to ignore the summary manner in which I was told of the decision a bare ten days before it was sealed and delivered by Council. . . .

> The treatment we have received will undermine the confidence of the laboratory in Council and also perhaps in me as Director, and it has left me with strong feelings of mistrust and alienation.[12]

Sydney Brenner was rarely coy about voicing his displeasure with MRC policies. On July 31, 1984, he penned a vitriolic missive to the Council with regard to its consideration of establishing Visiting Boards, which among other powers, would actively participate in determining financial allocations to MRC units.

> I am emphatically and totally opposed to this. It would be the most retrograde step ever taken by the Council, placing our budget totally under the control of University scientists who have no or little experience of the requirements of long-term, internationally competitive research, nor what it costs and how it should be organized. I have no doubts in my mind that this will lead to the end of the LMB and that, by comparison, the recommendations of the Phillips committee, which almost destroyed the Laboratory, will be like a vicar's tea party. . . .
>
> I spent the first part of my Directorship repairing the damage done to the Laboratory from the Phillips recommendations and other Council decisions. I will not spend the last presiding over the destruction of the Laboratory and, if I cannot prevent it, I shall leave the Council to commit suicide on its own.[13]

Tensions mounted concurrent to Brenner's frustration. Having devoted substantial time and energy to building on the LMB successes gained during the Perutz era, he was faced with colleagues that failed to share his vision for the scientific future of the laboratory. The central administration, in his view, was equally oblivious of the future of molecular biology. "I was initially very keen on Jim Gowans as Secretary of the MRC," Brenner stated. "He was a distinguished immunologist from Oxford and I was pleased that a real scientist was going to take the helm at the MRC. But things didn't turn out that way."

It is fair to state that, when times were good and resources flowed to the LMB in a steady stream, the director enjoyed a satisfying resonance with the MRC leadership. However, times changed, leadership at both organizations changed, and Brenner had a very different sort of relationship with Gowans and with his successor Dai Rees. Other issues fed into Brenner's increasingly disconsolate view of directing the LMB. The Thatcher administration keenly encouraged collaboration between the academic and private sectors, and in the mid-1980s sought to establish an MRC Collaborative Center (MRC CC), to be run jointly by the Mill Hill Laboratory in North London and the LMB. Expected to seek and acquire access to academic developments with commercial potential, the center was to promote technology transfer initiatives. Bronwen Loder was intimately involved in this program, but, from the LMB's point of view, it faltered over discord as to who would run the center. Eventually, the LMB was left completely out of the picture.

This background of continual frustration prompted Brenner to think increasingly about leaving the directorship and returning to his true passions—supervising his research laboratory and traveling the world, exchanging scientific news and ideas with his many colleagues.

As I look back on the period of being the director it was a mistake because it's not the sort of thing I really enjoy doing. I also realized post hoc that the gestalt of the MRC, indeed of molecular biology in general in the UK, had changed by the late 1970s. We were in a difficult period between the time of Asilomar and 1986. Scientists in the UK were nervous about the new recombinant DNA technology and people were milling around, uncertain about the future of this technology and uncertain about what new research directions should be adopted. I believed without a doubt that molecular biology was the future. In my own group we were cloning *C. elegans* muscle genes and I could see that even for structural studies people should learn how to clone genes and express recombinant proteins. I went so far as to inform people in the LMB that this technology was available in my laboratory and that anyone was welcome to send a postdoc or a graduate student to learn how to clone their favorite gene. As I recall only a single person ever did this.

this idea also. In 1985 when Gowans asked Brenner whether he wished to consider another term as director, he refused. Though scheduled to retire as director in 1987 at age 60, Brenner added that he would be happy to step down earlier if Gowans so wished. Indeed, in late October 1986, a mere month after the Loder saga, Gowans communicated this sentiment to members of the MRC Council.

> Dr. Sydney Brenner's current appointment as Director of the Laboratory of Molecular Biology in Cambridge formally comes to an end in 1987, the year in which he reaches the age of sixty. While it would be open to the Council to extend his appointment to age 65, the normal retirement age for members of the scientific staff, he has told me that he would like to give up the directorship in 1987.[4]

The news spread quickly. In November 1985 *Nature* carried the first of three editorial announcements concerning Brenner's future at the LMB.

> Britain's top job in molecular biology will become vacant in 1987, when Dr. Sydney Brenner plans to step down as director of the Medical Research Council's Laboratory of Molecular Biology at Cambridge. Brenner said last week that he plans to leave the laboratory at age 60, so as to be able to spend more time on his own research. He hopes to find himself a base elsewhere in Cambridge, and a small research group with which to pursue his present interests.[5]

There is little question that Brenner had grown weary of the directorship and its many undesirable trappings. The entrenched position of the senior scientific staff with regard to the future direction of the LMB, his disappointment at the collapse of the neuroscience enterprise, the dire state of research funding in the United Kingdom, and, last but by no means least, his intense desire to devote undivided attention to his research—all contributed to this state of mind. Bronwen Loder's abrupt dismissal was the last straw. "I could see the writing on the wall and when I decided that I no longer wanted to be director I urged the MRC to put a committee together to choose a successor. I strongly believed that the next director should come from outside the LMB—and I told them so." Gowans and the MRC Council shared this sentiment, at least initially. Indeed, the *Nature* editorial continued:

> The consequences of the vacancy at Cambridge will be considered by MRC at its meeting later in the month. The secretary, Sir James Gowans, said last week that all concerned are aware of the importance of the appointment, and that Brenner's successor will be sought on the international market.[5]

Anxious to ensure that his successor was someone of unquestionable scientific stature, Brenner prevailed on several individuals to offer their candidacy for the directorship, including cell biologist Martin Raff and the biochemist Alan Fersht—but without success. He was thus disappointed—even displeased—to learn that the selection committee, convened on June 19, 1986, unanimously agreed to offer the directorship to Aaron Klug, a long-standing LMB insider. Brenner, therefore, informed the MRC that he wished to resign the directorship a year earlier than planned—in 1986 instead of 1987.

Brenner was not officially required to retire as an active scientist until 1992 when he reached the age of 65. (Of course, he had no serious intention of ever doing so.) Thus, a few weeks after these events, he was considerably mollified when the MRC Council offered to establish a new research unit (the MRC Molecular Genetics Unit) under his direction. The council included the express condition that the new unit would be subject to review in 1990, and would close no later than September 1992 when Brenner must formally retire as an employee of the MRC.

There remained the contentious issue of where to establish the new unit. As we saw earlier, the MRC had a long-standing policy that required a unit director to physically vacate the unit upon retirement. Given his enormous stature in the United Kingdom, Brenner anticipated being exempt from this policy. After all, when Max Perutz retired as head of the LMB in 1977, he was not asked to leave the premises. This sensitive issue surfaced during a meeting with the MRC Council, but without satisfactory resolution. The very day of that meeting Brenner wrote to one of the Council members, the South African born Lewis Wolpert, a longtime colleague and friend.

Dear Lewis,

You asked me at the meeting today what I thought about staying on in the building after I relinquish the Directorship of the Laboratory. The Secretary has also raised this with me and I still don't know whether to consider the question flattering or insulting. My reply was lighthearted but I hope it communicated my view that I have taken all reasonable steps to separate myself from LMB. Since this issue has now been raised with me on several occasions I would like some assurance that it will not colour the decision whether or not the Council [will] grant me a Unit. . . .[6]

When copying this letter to MRC Secretary Gowans, Brenner added the provocative note: "I hope that this does not become a negotiating issue with my successor [Aaron Klug]."[7]

Dismayed at this unanticipated impasse, Brenner informed the MRC that he would find alternative space in which to work. He vacated the research laboratories he had occupied for so long at the LMB, moving temporarily into unused space in the LMB building that technically belonged to Cambridge University. As this was never intended to be a long-term solution, Brenner began casting around for a more permanent (and more suitable) location to house his new MRC Unit for Molecular Genetics. At the 11th hour, an opportunity came from Sir Keith Peters, an eminent British physician and the newly appointed Regius Professor of Physic at Cambridge University.[b] When asked to confirm that it was he who orchestrated Brenner's move to the Addenbrooke's Hospital, located across the street from the LMB, Peters was modest. "Nobody orchestrates Sydney," he laughed. "In fact he orchestrated me I think."[8] Peters' and Brenner's relationship dates back to the early 1980s when Peters was professor of medicine at the Hammersmith Postgraduate Medical School in London. An academically oriented physician, Peters had long been keen on promoting molecular biology in the practice of medicine. Indeed, he once convened a series of Saturday morning lectures given by distinguished scientists at the Hammersmith, some of which Brenner attended. Some time later, when Sydney's wife May required a professional opinion on a surgical matter, Brenner brought her to the Hammersmith. He used this opportunity to briefly serve as a visiting professor at the Hammersmith, dropping in on the various research groups and talking—especially to the younger set.

The Addenbrooke's Hospital was founded in 1766 on Trumpington Street, Cambridge, with seed money from the will of Dr John Addenbrooke, a fellow of St Catharine's College, Cambridge. In 1976, the Addenbrooke's was formally designated as the principal teaching hospital for Cambridge medical students under the direction of Lord John Butterfield. (Until 1976 Cambridge medical students undertook their preclinical education at the university, but were required to move to London for their clinical training.) In 1985 Peters was invited to succeed Butterfield as the Regius Professor of Physic at Cambridge. Delighted at the prospect of directing a teaching and training hospital in such close proximity to the LMB, Peters leaped at the opportunity to make good of this situation as soon as he heard of Brenner's need for laboratory space. "One of the primary reasons that I undertook

---

[b] The Regius Professorship of Physic is one of the oldest professorships at the University of Cambridge, founded by Henry VIII in 1540. Physic is an old word for medicine (and the root of the word physician), not physics.

this venture was because I knew that the LMB was right across the road and that Sydney was there," Peters related.[8]

> At one time I had the idea of trying to bring Sydney to the Hammersmith, not realizing then how deeply entrenched he was in the LMB and in King's College. So I went to Cambridge instead! So it wasn't as much a case of me rescuing Sydney from his dilemma of where to relocate as it was my very much wanting him to join the Addenbrooke's.[8]

With financial assistance from the MRC and the pharmaceutical giant Smith Kline Beckman, augmented with various funds squirreled away, including his Louis Jeantet award, Brenner was accommodated in renovated space in the Department of Medicine. The new Unit for Molecular Genetics was officially (and grandly) opened on May 10, 1989.

Brenner enjoyed free rein of his research program at the Addenbrooke's. In the interests of promoting research in the Department of Medicine, he agreed to mentor young physician-investigators eager to incorporate the "new biology" in their own work. "Sydney doesn't tolerate fools easily, so these youngsters found it rough going at first," Peters related. "But they were a stellar bunch and soon learned how to work with him. And even though Sydney makes jokes about having failed his final examination in medicine in South Africa, he knows a lot about clinical medicine."[8]

In truth, regardless of the fact that he was never a practicing physician, Brenner has a keen interest in forging links between the bench and the bedside—more important in his view, from the bedside to the bench. In fact, as early as 1979 he communicated his sentiments on this issue to Benno Schmidt, chairman of the U.S. President's Cancer Panel.

> It seems to me that we have still not yet solved the problem of how to establish the chain of continuity from the basic research laboratory to the practicing clinic. This is true in nearly all areas of medical research and the problem is compounded by what is almost a cultural antagonism between the pure and the applied. In a very small way we are trying to bridge this gap here in Cambridge by establishing a small tumour biology group in a laboratory devoted to fundamental research in molecular biology.[9]

Always skeptical of fads, Brenner is still guarded about the rallying cries for so-called translational medicine. He told an interviewer:

> Someone once asked me for a definition of translational research and I replied that it is the research that nobody wants to support. Those involved in basic research think it's applied and those who support applied research think it's too early and too risky. Translational research is based on the

assumption that there are smart people in basic research who are discovering important things and who will find some way of translating these to the treatment of human disease. This is called "from bench to bedside." I personally think that the focus should be the other way around. We should have "from bedside to bench" research by bringing more basic science directly into the clinic. We should also stop working exclusively on model organisms like mice and shift the emphasis to humans, viewing each individual patient as an experiment.[10]

In July 1986 *Nature* publicized Brenner's formal resignation from the directorship of the LMB.

> Dr. Sydney Brenner, director of the Medical Research Council (MRC) Laboratory of Molecular Biology in Cambridge, is soon to have a chance to give up administration and pursue his own personal research. This week the MRC announced that it had set up a new Molecular Genetics Unit which Brenner will head from 1 October this year until his retirement. The unit will be a small one, of 4 or 5 researchers.[11]

Brenner's research space in the Addenbrooke's Hospital was not ready for occupancy until 1988. In the interim he continued working in temporary space at the LMB, but was in the uncomfortable position of having to negotiate extended use of some of the facilities with the new director, Aaron Klug. Despite their common South African heritage, their contemporaneous education at South African universities, and their long-standing positions at the LMB, Brenner's and Klug's personalities are very different. With the added tensions surrounding Brenner's giving up the directorship, conversations between the pair took on a prickly tone. When Brenner met with Klug about using core LMB facilities and equipment, Brenner was informed that he should submit a formal application for use of any LMB facilities, as he was no longer automatically entitled to carte blanche access. Highly displeased with this directive, Brenner (who held the view that he certainly enjoyed a moral, if not a legal right to use the LMB facilities) summarily left Klug's office and notified him in writing of his intention to take this matter up with the MRC.

> Dear Aaron,
>
>   You said yesterday that you had agreed in principle to giving me access to certain facilities of the laboratory but that I would not have these "as a right." I have been thinking about what this right might be and it is now clear to me that I need to put any arrangements on a proper basis. There is no question of any dispensations on the part of the laboratory and I will

now, as part of my own negotiations, ask the [MRC] office to deal with this matter.[12]

Brenner's original agreement with the MRC was that support for his new unit would terminate at the end of September 1992, an arrangement explicitly stated in writing by MRC officer John Alwen in the summer of 1986.

> It is the Council's expectation that, because of its essentially ad hominem nature, the Unit will close on your retirement (or resignation as its Director) and that, as things stand at present, this will be in 1992 when you reach the normal retirement age for members of the Council's non-clinical scientific staff.[13]

Softening its position somewhat, the MRC directive went on to indicate that a special request for postponement of his retirement might be entertained at the appropriate time:

> While this will not preclude your asking in due course that exceptionally your own appointment and hence the Unit's tenure be extended, this would require a special case to Council."[13]

In early 1990 Brenner availed himself of this implied opportunity and requested postponement of his formal retirement from the MRC beyond his official retirement age of 65 (in 1992), citing recent exceptions to such hard and fast rules.

> My successor at LMB [Aaron Klug] got a 2-year extension of his appointment before he accepted the Directorship, on the grounds that it would adversely affect his research. Something similar is owed to me for the ten years I lost from my research, first as proleptic and then as official Director of the LMB.[14]

In March 1990 the MRC communicated a sympathetic, if somewhat guarded, opinion about this request to Council members, noting that Brenner's human genome studies in the new MRC Unit at the Addenbrooke's merited special consideration in this regard:

> In recent years only two exceptions have been made to the Council's policy on retirement age for scientific staff, both explicitly lined to the directorship of Units. ... The second, referred to by Dr. Brenner, was when Sir Aaron Klug asked for Council's agreement to a 2-year extension when he accepted the directorship of the LMB. ... Council would no doubt wish to consider the request very carefully since many directors of Council establishments

son (who retired from the Directorship in 2006) are scientifically active
boratory at the time of this writing.

formal retirement party was held for Brenner. "One day he was in the
and the next day he wasn't," Jonathan Karn stated. "It was as sudden as
However, Karn and fellow-postdoctoral fellow John Rogers took up a
llection to purchase a gift for their former mentor. After much delib-
and in deference to Brenner's long-standing interest in Japanese
and sword making, the pair presented him with a fine example of
form. Of course, as soon as the gift was in Brenner's hand, he drew
d from its scabbard and, waving it threateningly over his head, limped
on his one sound leg yelling in mock Japanese![19]

en asked about another rumor that circulates still in some quarters—
summarily dismissed the double Nobel Laureate Fred Sanger as soon
er reached retirement age—Brenner is emphatic:

t is absolutely not true. By that time the policy of people having to leave
ab once they were at retirement age was no longer a rigid one. So I asked
the year before he was due to retire (in 1983) whether he wanted to con-
e working at the LMB. I told him we'd love to have him remain, and that
so wished I'd move him administratively to the Director's Division, but
ouldn't have to move physically. But Fred adamantly stated that he abso-
ly wished to retire on his 65th birthday. He told me that he had long
ted to build a boat, and being a keen gardener he also wanted to tend
substantial English garden. So he literally put down his pipette, went
ie farewell party for him—and never returned.

could argue that they had not been able to devote as
research as they would have wished. In the present ca
additional argument that Dr. Brenner has played a ci
the Human Genome Project and that it would be in
that he should continue to be closely involved. This co
by extending his appointment.[15]

In the end, it was decreed that when Brenne
in 1992, he should "transfer to the Council's Exte
the contract under the Human Genome Mapping
be extended, but as an Outstation of the Resource
30, 1994.[16] In June 1991, a good year before the Un
was scheduled to close, MRC Secretary Dai Rees fo
for his services in a brief note—the very least c
outstanding and sustained contributions to science
during the preceding 35 years:

> It is ... my pleasure to write on behalf of the Cour
> warmly for the work you have done as Director of
> course your strenuous efforts and your achievem
> Genome Research.[17]

Brenner received a more direct (and distir
reminder about the imminent approach of Septei
MRC personnel officer. Presumably striving to be c
contacted Brenner concerning the impending tern
personnel in the Unit of Molecular Genetics. But
the MRC establishment was now worn out. "I resent
randum," he wrote angrily. "I do not see any cause for
administration would like to dispatch me and my g

Keen to have the last word about this phase of
the MRC, Brenner posted a note under a plaque that
of the MRC Unit in Molecular Genetics on May 10,
and closed on September 30, 1992." With Keith Pete
retained his laboratory at the Addenbrooke's for and
ing his research efforts with the private funds award
the title of Honorary Professor of Genetic Medicine
Cambridge University, which was bestowed on him

Ironically, the MRC no longer requires that I
facility they worked in following retirement. Both
quished the Directorship of the LMB in 1996) an

PART 5

*Life Outside
the Laboratory*

22

Finding New Opening Games

Not that bloody woman again!

WHEN HE ASSUMED FORMAL DIRECTORSHIP OF THE LMB, without relinquishing or diminishing his research programs, Brenner's plate was more than full; it was overflowing. Still, he managed to eke out time and energy for other new and substantial responsibilities—new games, one might say. Primary among these was his ascendancy to become the Editor-in-Chief of the *Journal of Molecular Biology* (JMB) in 1987, a position he retained until 1990. Founded by John Kendrew in 1959, the JMB was for many years one of the premier journals in biology. Indeed, in the period between 1959 and 1975, the journal featured many of the fundamental discoveries in molecular biology.[1]

In time the journal became less of a priority to Kendrew, and its reputation slowly declined. New leadership was clearly needed. However, Kendrew was reluctant to relinquish control of the journal, "which he saw as his own baby."[1] The result was a progressive decline in the journal as a vehicle for publishing cutting edge work. Brenner, with characteristic energy, met the challenge of changing the thrust of the periodical and saved the situation. His vigorous use of scientific contacts around the world soon revitalized the quality of the journal, which slowly recovered lost ground. In 1987 he replaced Kendrew as Editor-in-Chief.[1]

Once or twice a week, Gillian Harris, Brenner's editorial assistant, would enter the laboratory laden with a stack of 15 to 20 manila folders containing manuscripts that needed his attention. Jonathan Karn relates that whenever Sydney would see her walking down the corridor, he would shout, "Not that bloody woman again!" Nonetheless, he would immediately drop whatever he was doing and rush off to attend to JMB matters. Gillian, long accustomed to this curious salutation, would smile charmingly and encouragingly proclaim, "Not too much this week," despite the fact that the piles of manuscripts were never ending.[2]

As with everything else to which he commits, Brenner took his editing responsibilities seriously, filtering out mediocre papers and carefully overseeing the peer-review process of those that, in his view, merited publication. Brenner's notation of "Bumph" and "more bumph" on submitted manuscripts became in Gillian's translation: "Your paper was considered at a meeting of the Cambridge editors. I regret to inform you that the board felt that it falls outside the scope of the journal's editorial policy and would be better suited to a more specialized journal." And Brenner's protest of "Oh no, not again!" was rendered as "We regret that owing to the limitations of space we are unable to consider your manuscript." If Brenner was convinced that a contribution, unappreciated by the referees, had merit, he would go to great lengths to accept it, occasionally overruling negative reports with which he disagreed. This was particularly true of papers reporting innovative techniques that were criticized because their applications were judged to be too limited.

In 1989, Academic Press, publisher of the JMB, invited Brenner to identify the most significant papers published in the journal during the preceding 30 years. These were reprinted in a special volume entitled *Molecular Biology: A Selection of Papers.*[3] After repeatedly culling his list of candidates, Brenner selected 38, beginning with a famous contribution by Pardee, Jacob, and Monod published in the inaugural issue. In his preface Brenner referred to the group of 38 papers as representing steps on the continuing road of molecular biology, emphasizing how they had led to an understanding of how the genetic code is deciphered to yield proteins and how these, in turn, interact to render the regulated behavior of cells. He also stressed the historical value of the collection, especially for students, who, in the late 1980s, were already becoming seduced by the proliferation of ready-made kits for performing experiments. "Many ideas, which today we find simple and obvious, only became established by complex and difficult argument as people struggled to extract themselves from the rigid mould of contemporary thinking. And many experiments that today are carried out with kits ordered by telephone, were more than heroic acts at the limit of existing technology."

Brenner relinquished the position of Editor-in-Chief of the JMB on June 1, 1990. In response to a congratulatory letter from a colleague lamenting his retirement from this position, Brenner replied: "Thank you for your kind note. After 30 years they even let murderers out of jail!"

In 1959, a mere three years after arriving in Cambridge, Brenner was elected a Fellow of King's College. He was provided with furnished rooms "with an interesting view" where he could escape to work undisturbed— and sleep over if he wished. He enjoyed the privilege of eating at High Table, which he indulged to the fullest.

In earlier times the dining halls were a major aspect of Cambridge college life. Gowned students would solemnly rise as the College Fellows entered and walked the length of the great Hall to the High Table. Latin grace would be delivered by a scholar, following which the fellows would wine and dine. Undergraduates, seated in the body of the hall, would have to contend with food of indifferent quality.

Brenner often dined at High Table (and does still when the occasion lends itself), delighting in the intellectual camaraderie afforded by fellows outside the life sciences. For many years College Fellows were assigned seats in order of seniority. But this formality was abandoned before his day. Brenner enthusiastically availed himself of opportunities to mingle freely with peers from other academic disciplines. He cultivated a close relationship with the distinguished English philosopher and former University Provost Bernard Williams,[a] and for 25 years he sat on the committee that elects new fellows. "Sydney was a Cambridge College person to an extraordinary degree—in a way that's hard for me to even get my head around," Brenner's friend and medical colleague Keith Peters stated. "He was very deeply embedded in the King's College scene."[4]

The gradual erosion of ritual and stodgy formality at High Table did not excuse what some might consider abject table manners. Brenner recalls (with sardonic amusement) an occasion when a silver tray laden with a single bunch of grapes was passed around the table. Vast quantities of grapes being no novelty to a South African, Brenner unhesitatingly set the entire bunch on his plate. It took no more than a second or two for him to register the sea of disapproving faces. "As a Fellow," Brenner explained, "I got to participate in a new level of intellectual life at Cambridge. I met a lot of interesting people at High Table; other scientists and fellows from other disciplines at Cambridge. In fact with time May and I spent less and less social time with people from

---

[a] Sir Bernard Williams (1929–2003) has been described as the most important British moral philosopher of his time. He was renowned for being sharp in discussion, with Oxford philosopher Gilbert Ryle once saying of him that he "understands what you're going to say better than you understand it yourself, and sees all the possible objections to it, all the possible answers to all the possible objections, before you've got to the end of your sentence." See http://en.wikipedia.org/wiki/Bernard_Williams.

the lab. One saw enough of these people at work." May Brenner also cultivated a group of friends through her work as a psychologist. By the early 1960s the Brenners were very much part of the Cambridge scene, and by the mid-1960s, "there was no question in my mind about moving, though I had been offered various jobs elsewhere."

It was at King's that Brenner nurtured his well-known taste for fine wines. When the College was bankrupt during the 19th century, the fellows dipped deeply into their own pockets and heroically rescued the wine cellar. Since then the facility has technically belonged to the fellows, and the college is required to sell wine to them at cost. Brenner continues to avail himself of this opportunity, and, in fact, stores much of his personal wine in the spacious King's College cellars. Over the years his knowledge of wines evolved to that of an accomplished connoisseur, and his correspondence includes many letters of thanks to individuals and organizations that gifted him special wines over the years. "Before I discovered fine wines I drank Commando brandy," he stated in reference to a cheap but potent South African alcoholic brew much favored by medical students with little money to burn. "But at King's I discovered the joys of fine wines."

Much to their satisfaction the scientific staff of the LMB had no formal teaching obligations to Cambridge students. Eager to attract bright students to the life sciences, however, and keen to awaken their interest in the LMB and its illustrious reputation, Brenner, and sometimes others, delivered lectures for the department of biochemistry. Additionally, Brenner occasionally organized seminars for interested students, a useful mechanism for identifying promising students who might be lured to careers in molecular biology.

When Philip Goelet, a graduate student in biochemistry at Cambridge in the early 1980s, became bored with his research on polysaccharides, he began casting idly around for a more exciting field in which to pursue his doctoral training. While listening to a seminar by Max Perutz, he became increasingly energized and was delighted when Perutz extended an open invitation to the class to drop into the LMB if they wanted to learn more about its research. Goelet did so and, eventually, appropriated Brenner as his thesis mentor. Goelet related:

> Sydney taught a graduate course at Cambridge while I was a student at the LMB. He would walk into the lecture theater and ask: "Well, what do you want to talk about today?" Someone would mention a recent paper they knew of and Sydney would immediately say, "Yes, that's a good choice. Let's talk about that paper. Do you remember this figure... ?" And he'd proceed to accurately reproduce a graph from the paper in question on the blackboard. Of course this was extremely impressive to us.[5]

As well as directing the LMB, managing his ever-widening research program, running the *Journal of Molecular Biology*, occasionally teaching, and traveling far and wide, Brenner served as a consultant to Biotechnology Investments Limited (BIL), one of the many venture capital cogs in the vast Rothschild Asset Management wheel in the United Kingdom. As we saw in Chapter 16, Victor Rothschild and Brenner were scientific colleagues at Cambridge and had cultivated a warm personal relationship. Rothschild was a compelling individual. "The records of his accomplishments and the diversity of his interests were enough for a dozen men," biographer Kenneth Rose wrote.[6] "He was equally at ease in those supposedly rival cultures, science and the arts." An accomplished zoologist (he was elected a Fellow of the Royal Society for his research contributions on fertilization) and a thoroughly eclectic personality, Rothschild assembled one of the finest collections of 18th-century English books, bindings, and manuscripts in private hands.[6]

Despite his peripheral association with members of the infamous Cambridge spy ring, Rothschild served as head of countersabotage in the British Army during World War II. In fact, he received the George Medal for discovering and defusing an explosive device hidden in a cargo of Spanish onions bound for Britain. Rothschild was also entrusted with the responsibility of examining the safety of gifts of food, drink, and cigars to Winston Churchill.[6] He spent a decade with Royal Dutch Shell, rising through the hierarchy to become coordinator of research for the group worldwide. Prime Minister Margaret Thatcher often consulted him while he was Chairman of N.M. Rothschild and Sons. Rothschild was a formidable cricketer, "intrepidly facing the bodyline bowling of Harold Larwood and earning an obituary tribute in Wisden Cricketers' Almanack sixty years later."[6] He was also an adept jazz pianist and is said to have set records "scorching between Cambridge and Hyde Park Corner in Mercedes and Bugatti."[6]

When public interest in the potential commercial gains of recombinant DNA technology (biotechnology) gained serious traction in the early 1980s, Rothschild founded Biotechnology Investments Limited (BIL). He lost little time in securing Brenner as his senior biotech consultant. In early 1981 he wrote to Brenner outlining his expectations:

Dear Sydney,

As you know we, N.M. Rothschild & Sons Limited, intend to set up an organization which will enable people to invest in biotechnology. We have

in mind both quoted and unquoted companies or even, possibly, organizations. We are most anxious for you to become a Consultant to our organization, whose name and structure have not yet been completely determined,
and we would like to offer you $— per year plus expenses to undertake
this job. We envisage two classes of activity in which we would rely on
you for guidance. First, we shall from time to time send you details of proposed investments and ask you one or other or both of two questions about
them: firstly, do you recognize any of the people concerned with the undertaking in question and, if so, what do you think of them from a scientific
point of view? And what do you think of the subject or subjects on which
the undertaking is working?

Secondly, we would hope that you would periodically let us know of
undertakings which you yourself think worthy of support, either by virtue
of the people involved or their subjects.[7]

So began a formal association with BIL that endured until Rothschild's
death in 1994. Correspondence from Rothschild to Brenner was frequent
and typically succinct, sometimes comprising nothing more than a single
line, such as, "Come to lunch on Saturday." In contrast, Brenner's reports
(invariably hand-written) were lengthy, thorough, and unusually incisive.
He reported in detail on the strengths and weaknesses of biotech companies
in which BIL was interested in investing, reducing the science to intelligible
language and often volunteering investment opinions.

At this time working for the private sector was frequently frowned upon
in academia as prostituting one's scientific skills. Brenner was sensitive to the
reactions of some of his scientific colleagues, especially in light of his position
as Director of the LMB. In early 1984 he wrote to David Leathers of N.M. Rothschild and Sons, requesting that his name not be included in the BIL prospectus
in case this was picked up by the public press. "I am in a special and very delicate position because I have to monitor, and occasionally censure, the activities of my colleagues, and I therefore need to . . . leave no possibility open for
activities to be criticized," Brenner wrote.[8] However, he was quick to assuage
concerns at BIL that he might discontinue his consultative activities for this
reason. "As I said over the telephone, I shall make myself personally responsible for the $10,000,000 you will fail to collect by not using my name, by
remaining a consultant for you for the next 500 years." He concluded his missive with the rejoinder: "Yours forever."[8]

Victor Rothschild died of a heart attack at St James's Place, London (the
headquarters of his business), on 20 March 1990, and was buried near the
tomb of his great-grandfather N. M. Rothschild.

23

Mounting a Human Genome Project

It was sheer murder

SYDNEY BRENNER IS AN EXTRAORDINARILY DILIGENT CORRESPONDENT. Letters are sometimes not dealt with for weeks—even for months. However, they are all ultimately answered, frequently in his impeccably hand-written script. At one extreme of the diligence scale, he once received a letter from a company that manufactured laboratory equipment. Addressed to "Dear Scientist" and obviously mailed in bulk, the letter requested the completion of a market survey on procedures used for DNA binding. Nowadays even lowly graduate students immediately discard such correspondences. Brenner responded with a letter apologizing for not completing the survey.

On the other extreme of timeliness, a letter dated June 1991 begins:

Dear . . . ,

I feel ashamed at having not replied to your letter of almost 2 years ago, but I lost it and it only surfaced now.

One is struck by the number of communications—some quite detailed—that Brenner has written to people he hardly knew (some not at all), who sought advice or who rendered opinions (frequently zany) on some scientific matter or other. Upon receiving a letter from a 17-year-old Irish student in the lower sixth form who was contemplating a career in molecular biology, Brenner—after graciously apologizing for a one-month delay in responding—took the time to offer the young man the following advice.

25 June, 1970

Dear Mr. Turner,

I am not at all certain that I can advise you what to do at the undergraduate level, since most of my experience is with postgraduates. What is clear is that a strong background in chemistry is essential to do molecular biology, since most modern molecular biology is biochemical in nature. A pure physics degree restricts one strongly to the choice of field in biology one can enter, although it is a good background for doing work on more structural aspects. Of course one can take rather general courses in biology, many of which emphasize the molecular aspects. You would do well to write to people about this. I can suggest Edinburgh as one possible place and Leicester as another. In these Universities courses are given which combine the genetic and biochemical approach. You may also be interested to know that at Warwick University it is possible to do a degree in molecular sciences that is mostly physics and chemistry, taking courses in molecular biology and biochemistry at the same time. Cambridge, of course, has a very flexible natural science tripos and here it is possible to study a combination of chemistry and biology, which is a very good background for molecular biology.

Yours sincerely,
S. Brenner[1]

Brenner kept meticulous records of essentially everything that crossed his desk—he is a historian's dream come true. Many scientists diligently archive correspondence received, but such collections frequently lack copies of correspondence sent to others. Brenner kept copies of all his communications. This habit may reflect nothing more than his association with well-organized and conscientious assistants. But his sensitivity to the importance of historical records is evident in a relatively recent letter to *Nature*, written with Nobel Laureate Richard Roberts, in which he urged the molecular biological community to preserve their records, especially in this paperless era of electronic mailing.

> "Let's not wait until memories have faded and papers have been discarded at the end of a career before deciding to save our heritage," the pair wrote. "Future historians of science and social science should not have to look back and wonder how it was possible that we discarded the records of our lives in science."[2]

A much sought-after scientific celebrity, Brenner was (and still is) pursued for all manner of activities, especially participation in scientific meetings, and frequently as a keynote speaker. He declines many of these invitations, but in truth he thrives on the camaraderie and the lively verbal sparring that distinguish the best scientific meetings, and is never shy to offer scientific criticism and advice. Brenner rarely uses notes, relying on his exceptional memory and an astonishing gift for organizing thoughts while on his feet. He thus typically lectures without the use of visual aids, and when these are essential, he prefers the flexibility of overhead projections to slides. PowerPoint presentations are an anathema to Brenner.

His long-time friend from South Africa, Phillip Tobias, tells of an occasion in 2003 when Brenner visited South Africa to receive an honorary doctorate at his alma mater, The University of the Witwatersrand. This event required that he offer the commencement address to a class of graduating medical students, who, together with their many relatives and friends, packed an enormous lecture hall on the university campus. The night before commencement he and Tobias dined out. Afterward Brenner, who presumably had paid no attention to the request for a commencement address, casually asked Tobias whether he would be expected to say anything at the graduation ceremony the following day. Somewhat shocked at the late hour to be asking such a question, Tobias politely informed him that it was customary for the recipient of an honorary doctorate to address the graduating students. "The next day when he got up to speak he placed a few folded pages of notes on the podium and proceeded to speak, never once unfolding the notes. He could ad lib and improvise like no one I know," Tobias related.[3] The extensive Sydney Brenner archive housed at the Cold Spring Harbor Laboratory includes a draft of his address on the evening of the Nobel Prize awards. His remarks are written in longhand—on stationery from the Grand Hotel, Stockholm.

Brenner refuses to write manuscripts for the published proceedings of meetings and symposia, believing that these summaries are more likely to swell the coffers of publishing houses than to inform the community about scientific progress in a timely manner, further compromised by inevitable publication delays. He once wrote to Maurice Wilkins:

> In your invitation you have included the ultimate deterrent to my participation in the meeting, that is the demand for a manuscript. I just cannot find time to sit down and get involved in writing an essay on this subject for publication. I am sorry.[4]

This attitude extends to most invitations to write review articles. However, Brenner is not averse to what he considers purposeful writing, and he has provided many editorials, commentaries, and correspondence published in leading scientific journals. His numerous opinion articles and book reviews, occasionally marked with characteristic derision accompanying damning criticism, always impart ideas that provoke serious reflection.

Brenner is tolerant of textbooks, which he believes can be important instruments for learning. While he often alludes to the worthiness of book topics, both scientific and otherwise, he has not (yet) given in to the demands of writing a book himself. In a typical response to a plea for a book from a major publisher, he wrote,

> I have no fixed plans to write a book at the moment, only rather vague intentions. Eventually is the operative word.[5]

Brenner devours the scientific literature. "Francis Crick once told me that he had stopped reading the literature in molecular biology because he was sure that if anything important happened, somebody would tell him about it," Brenner stated. "The trouble is that nowadays he would have an endless queue outside his door and would listen to nothing else."

Brenner's engagement with the scientific community is intense and impressive—he is inclined to agree to most requests for participation in scientific activities that he considers useful. His involvement in meetings, committees, and the boards of several scientific entities is balanced with his efforts in the development of academic activities in other institutions, including the evaluation of young scientists being considered for academic advancement at places far removed from his immediate spheres of interest. Mindful of his intensely busy schedule, meeting planners sometimes seek his commitment years in advance. Such long-term planning, however, can backfire if more urgent (or interesting) opportunities arise in the interim. Having once accepted an invitation to speak at a meeting a full two years hence, Brenner apologetically informed the organizers of competing demands on his time a month prior to the meeting:

> I am sorry to have to break a long-standing arrangement, but you can now understand why I dislike making commitments so far in advance.[6]

As we shall see in the next several chapters, Brenner became extensively involved in scientific affairs that took him around the world—to Singapore and Japan, and, more recently, the United States. Aside from the strenuous intellectual obligation that these activities demand, the travel is arduous. Despite the comforts of first- or business-class travel and limousine pick-ups, Brenner's travel schedule, even in his later years, is daunting. His pocket diary for 1996, when he was 69 years old, documents no fewer than 113 trips by air.

Recall that Brenner has had to use a cane much of the time, but there are other physical challenges as well. He also suffers from chronic emphysema, presumably the aftermath of chain-smoking for many years, which is frequently aggravated by bouts of allergic asthma, especially in the winter months. In 1994 he was treated for colon cancer, and, at the time of this writing, is recovering from painful tendonitis affecting both wrists.

The history of the Human Genome Project has been extensively documented in many scholarly books and articles and is not recounted here. But a brief review of this momentous venture will underscore Brenner's role in promoting a human genome project in the United Kingdom, an effort that—similar to the launch of recombinant DNA research—was mired in politics and competing interests. The conflict fundamentally centered on the value and utility of documenting the entire sequence of the 3×10^9 nucleotides of the human genome. On one hand were those who insisted that this gargantuan and expensive effort would surely yield vital information about gene function in health and disease. On the other hand were those who considered it an abject waste of financial resources, time, and energy that could be more gainfully applied to other areas of biomedical research. The best work, the mantra went, came from investigator-initiated studies in small labs, not from some massive goal-driven effort.[7]

In the United States the debate occupied much of the mid-1980s and the early 1990s. During this time the U.S. Department of Energy and the National Institutes of Health (NIH) competed for control of this gigantic effort, one that has been compared to that of putting a man on the moon. In late 1998, the NIH established an Office for Human Genome Research, which was subsequently elevated to the National Center for Human Genome Research, led by Jim Watson. However, when former NIH Director Bernadine Healy approved the patenting of cloned genes against Watson's wishes, he resigned his position.

Brenner related:

> I spent a lot of time trying to negotiate a human genome program in the United Kingdom. We had tremendous difficulty persuading influential funding bodies in the UK to think big about the genome. It was sheer murder! The real issue was that it had to satisfy some kind of threshold of new money so that it wouldn't be taking funds away from individual scientists.

Otherwise there would be a lot of screaming and shouting. That's really what all the politics was about. Where can we get new money?

Brenner and the British geneticist Walter Bodmer, then director of the Imperial Cancer Research Fund (ICRF), devoted countless hours lobbying for an independent U.K. genome project capable of claiming a significant stake in the race to sequence mankind's DNA. Bodmer devoted modest resources from the ICRF, and Brenner was responsible for the release of about 11 million pounds for genome research from the British government. Like their colleagues across the Atlantic, however, Brenner and Bodmer knew full well that a robust genome project would require far greater resources. Ultimately, they had to concede to the reality that the United States was going to be the only big winner in this race. "The multiple token financial gestures put the airplane on the runway," Keith Peters remarked. "But lift-off didn't occur until the Wellcome Trust emerged onto the scene."[8] Arthur Gibbs, chairman of the Wellcome Foundation, had the foresight to appreciate that converting the foundation to a private entity that could invest on a much broader basis would yield far greater financial dividends for biomedical research in the United Kingdom. Therefore, Gibbs created the Wellcome Trust, a veritable godsend for biomedical research in the United Kingdom.

Brenner crafted his own approach to genome research. His goal was not to sequence the entire human genome with its extensive repertoire of noncoding DNA, but to focus on the coding elements. "Because at the end of the day that is what everyone wanted." Brenner chanced on the ingenious notion of identifying coding elements by isolating mRNAs and converting them to single-stranded DNA copies—so-called complementary DNA (cDNA)—that would also facilitate mapping the genome.[a] He announced this idea at one of the early meetings, convened in Santa Fe, New Mexico, in 1986, to consider sequencing the human genome. "In my view this would have been functionally more interesting and given us access to all the coding elements much sooner," he stated.

"Wally Gilbert once asked me how I would like to contribute to the human genome project," Brenner stated. "I said to him: 'Wally, if I can characterize

---

[a] Craig Venter, the entrepreneurial DNA sequencer and founder of the company Celera, heard about Brenner's ideas around the time that he was independently pursuing the same notion. He referred to complementary mRNA in this context as EST's (expressed sequenced tags). The pair apparently agreed to copublish papers on the topic, but proprietary issues prevented Brenner from exchanging such data at that time. Venter JC. 2007. *A life decoded: My genome: My life*, p. 126. Viking Penguin, New York.

100 human genes in my lifetime I'll be content. Why waste one's time on 1000 genes?' " When Gilbert suggested that joining the human genome project would in time give him access to all the genes he wanted, Brenner responded: "You're like the guy who comes to Christopher Columbus and asks: 'Why are you messing around with wooden ships and Spanish sailors? Why don't you wait 500 years and you'll be able to get a cheap air flight directly to America?"

Brenner's focus on cloning and mapping human cDNAs never translated to substantial funding for such an initiative. The weight of scientific opinion shifted in favor of sequencing the entire 3 billion base pairs in the human genome. Ultimately, two teams competed for primary credit for this grand undertaking: one led by Francis Collins at the NIH, the other by the entrepreneurial Craig Venter, who founded the private company, Celera. The ensuing race to complete the sequence is one of many fascinating chapters in the history of the human genome project. At the close of the 20th century, President Bill Clinton declared this quest a tie. Summoning both Venter and Collins, he asked them to put aside petty competitive interests and to join him in a televised announcement revealing that the draft sequence of the three billion base pairs in the human genome was, for all practical purposes, now complete. A simultaneous television broadcast, featuring Prime Minister Tony Blair and John Sulston, recognized the more modest British contribution to this scientific milestone.

Meanwhile, Brenner made his first serious foray into genome sequencing with an exotic species of fish, *Takifugu rubripes*, colloquially known as fugu, or puffer fish. As we saw in Chapter 20, although this work began in his Addenbrooke's hospital laboratory, Brenner's interest in fish genomes dates back to the 1970s when he spent several summers at the Woods Hole Marine Biology Laboratory. A frequent patron of science libraries, Brenner used much of his free time exploring the holdings of the Marine Biology Library. While perusing issues of *The American Naturalist* (which he took pains to point out was not a nudist magazine), he came across an article[b] that documented the DNA content of several fish species. Brenner noted that some had a DNA content about one eighth that of man:

---

[b] Hinegardner R, Rosen DE. 1972. Cellular DNA content and the evolution of teleostean fishes. *Am Nat* **106:** 621–644.

> Now at the time if you asked anybody: "What do you think about this fish with eight times less DNA than man?" The answer would be: "Clearly that's all the DNA the fish deserve! I'm at least eight times more complicated than this fish."

Brenner had no firm idea of what DNA content meant, as the world had just discovered repetitive DNA. But a paper published in 1983 demonstrated that these fish genomes contained very little repetitive DNA. So he tucked this fact away, deciding that fish genomes would be interesting to examine more closely some day.

Also called blowfish, swellfish, globefish, balloonfish, or bubblefish, the puffer fish—so called because of their ability to inflate themselves to several times their normal size by swallowing water or air when threatened—are in the family of *Tetraodontidae*, in the order *Tetraodontiformes*. Tetradon connotes the four large teeth fused into upper and lower plates that are used for crushing the shells of crustaceans and mollusks, their natural prey. Brenner searched extensively to acquire some of these exotic fish species, but he was genuinely concerned about the unreliability of natural sources. His great worry was that pollution or some other catastrophe would render them extinct, in danger of extinction, or more likely, inaccessible because of stringent regulation. So he settled on a fish that is cultivated for food, the Japanese puffer fish, fugu.

> When people asked me, "Why are you interested in this?" I replied, "Let me give you a little talk on junk!" . . . Scientists were then divided into three classes. There were the physicists, who believed that organisms didn't have enough DNA to do what physicists thought were very complicated things; there were the smart molecular biologists, who thought that organisms had exactly the right amount of DNA; and there was a group, including me, that was really concerned about what this extra DNA was there for.

In the vernacular of molecular biology, noncoding DNA is frequently referred to as "junk,"[c] reflecting the notion that it has no specific function. Brenner, however, believes otherwise. "There is junk—which is rubbish that is kept, and there is garbage—which is rubbish that is thrown out. Most languages use two words to specify this distinction. In Japanese, *gomi* is rubbish you throw out, while *garakuta* is junk that you keep. Of course if there is rubbish

---

[c] The phrase "junk DNA," coined by Susumu Ohno (Ohno S. 1972. So much 'junk' DNA in our genome. *Brookhaven Symp Biol* **23:** 366–370) initially referred exclusively to repetitive DNA, but was soon used to describe all categories of noncoding DNA.

in your genome it must be junk because if it were garbage it wouldn't be there."

Fugu is perhaps best known in Japan and Korea as a dangerous delicacy; the eyes and internal organs of most puffer fish are highly toxic. In polite Japanese society, the host tastes the prepared fugu before it is served to guests, and licensed Japanese restaurants employ trained chefs expert in handling and preparing the fish. The puffer fish toxin (tetrodotoxin or anhydro-tetrodotoxin-4-epitetrodotoxin), generated by bacteria in their food, is an exceptionally lethal neurotoxin that causes rapid respiratory paralysis. It is estimated that a single puffer fish contains enough poison to kill 30 adult humans. According to rumor, experienced fugu chefs in upscale Japanese restaurants will, if requested, leave sufficient poison in the fish to provide adventurous diners a touch of excitement—numbness and tingling of the lips, perhaps even the shortness of breath that signals the onset of respiratory paralysis. True connoisseurs purportedly enjoy hot sake mixed with fugu testes, one of its most lethal organs. This drink apparently generates a "buzz"— and supposedly acts as an aphrodisiac. Puffer fish are themselves resistant to the neurotoxin as the result of an evolutionary adaptation, a mutation in a gene encoding a protein in the sodium channel pump in their cell membranes.

Brenner mounted the fugu project in 1989, primarily through the efforts of Greg Elgar, a bright and enthusiastic physician-turned-scientist; Sam Aparicio, an enterprising Canadian postdoctoral fellow; and an Indian scientist from Singapore, Byrappa Venkatesh. According to Brenner, "The first thing we wanted to know is whether or not this fish has the same number of genes as we do. Because if it had only one-eighth the number of genes it obviously wasn't very interesting." This fundamental goal was achieved by using statistical genomics. The team sequenced six hundred random pieces of DNA and asked how many known vertebrate genes could be identified. They calculated the abundance of these genes in the fugu genome and demonstrated that they were eight times more enriched in the fugu genome compared to the vertebrate genome. Thus, whereas fugu and humans have more or less the same number of genes, the fugu genome is much more compact. Brenner explains:

> I like to call the fugu genome the "discount genome," because you get a ninety percent discount on sequencing it. So having enhanced the sequencing industry about ten-fold, I'd fulfilled the important technical requirement that everybody stated one should strive for—a ten-fold gain in technology every five years. And I'd done that in a few months by once again choosing the right organism.

"Let's assume that certain groups of genes have stayed together through-out the period of about half a billion years of evolution from fish to man," Brenner elaborated.

> It's relatively easy to map a human gene—let's say a gene for breast cancer. But that only narrows its location to a piece of DNA that's about a million base pairs long. But it's enormously tedious and time-consuming to sequence through a million base pairs. So we were saying that since a gene in man is about 50,000 base pairs long on average whereas a gene in fugu is about 6000 base pairs long, one would only have to sequence a hun-dred thousand, or a hundred and twenty-five thousand base pairs of fugu DNA to know the sequence of that gene.

Brenner reasoned that this important distinction would enable one both to locate human genes and to characterize unusually large human genes. Con-sider, for example, the size of the dystrophin gene affected in the disease Duchenne's muscular dystrophy. The gene is 2.8 megabases in length, whereas its fugu homolog is reduced in size by a factor of 20.

Brenner was persuaded by the benefits of the fugu genome project through the pursuit of what is now called comparative or evolutionary genomics.

> Let's suppose that I were to start a new genetic program and I wrote a grant saying, "I want to get a fish in the lab and I want to make mutants and I want to turn it into a man." Because that's the really interesting thing: how fish became men in the course of evolution. No one would give me a penny of course. But the fact is that this experiment has already been done for us by nature. All one has to do is take a fish gene of interest and ask whether that gene functions properly in man—or a mouse. From the point of view of physiology and anatomy it can be argued that the mouse must have "fishy" parts in its genes, plus something that got added on later in evolution. The immunology will still be "fishy." But the lungs certainly won't be because they aren't present in fish. So if I find a gene that is expressed in the lungs of a mouse and I find the same gene in a fish, I can ask whether that fish gene is functional in the lungs of a mouse. If it isn't I should exam-ine the mouse gene to see what happened to it during evolution. Have other parts been added on? Has it moved to another place perhaps where it is under different regulation?

But if the fish gene is, in fact, functional in the mouse, Brenner reasoned, one can reasonably conclude that the gene was unaltered during evolution. That is, if one constructs two animals that are absolutely identical, except that one has a piece of DNA from the mouse and the other carries the homologous DNA from a fish, everything that is common in their sequence must be critical for the gene's function. It is, of course, imperative in such analyses to restrict comparisons to extended evolutionary distances, in this case, that between fish and mice.

> We contain "mousy" features because we evolved from something that also gave rise to mice. But the evolutionary distance between mice and fish is sufficiently far that we should be able to distinguish between rubbish (DNA sequences that have been discarded) and junk (DNA sequences that have been kept even though they are noncoding.) I believe that this sort of experimental approach will enable one to work out the genetics of evolution and to decipher regulatory mechanisms in complex organisms.

Indeed, comparative genomics occupies much of Brenner's contemporary research focus, and such focus requires little more than a computer—and the patience to pour over millions of iterations of the letters A, C, T, and G.

When formally announcing fugu's "discount genome" in *Nature* in 1993, Brenner and his colleagues had no reservations in referring to it as "the best model genome for the discovery of human genes."

> Small "model" genomes such as those of *Escherichia coli* and yeast, or even those of *Caenorhabditis elegans* and Drosophila . . . may throw light on the genetic specification of more complex functions. However, vertebrates differ [from invertebrates] in their morphology and development, so the ideal model would be a vertebrate genome of minimum size and complexity but with maximum homology to the human genome.[9]

Talk is cheap when it comes to sequencing genomes, but acquiring the considerable financial resources required to translate intentions into reality is altogether another matter. A grant proposal to the NIH met with unfavorable reviews. Support from the Sanger Center was also not forthcoming—much to Brenner's chagrin. He commented that he was "interviewed by a committee of former postdocs!" One hard reality to contend with then was that the primary focus of funding agencies (including the Sanger Center) was the *human* genome. This source of intense frustration to Brenner prompted disparaging comments about the "genome politburo" that decided which genomes were going to be sequenced and which were not. "Unfortunately they simply didn't see the importance of comparative genomics then."

It was not until 2000 that Brenner was given financial support from Singapore, and, more important, from the Joint Genome Institute (JGI) of the U.S.

Department of Energy. The JGI is a consortium of three U.S. Department of Energy national laboratories—Lawrence Berkeley, Lawrence Livermore, and Los Alamos—which selected the puffer fish as a genome sequencing project. In November 2000, the International Fugu Genome Consortium was formally organized, representing one of the largest international genome-sequencing ventures since the historic Human Genome Project.

Over a period of about nine months, from late-2000 to mid-2001, the consortium steadily generated sequences. However, it soon became obvious to Brenner and his colleagues that this effort was far too slow. They, therefore, contracted with a private company to obtain sequences more rapidly, but the company in question was plagued by technical and operational problems and the contract was annulled.[10] In the interim others had become aware that sequencing the puffer fish genome could provide a useful shortcut to human genes. At one point Hans Lehrach, from the Vertebrate Genomics Group at the Max Planck Institute in Germany, proposed to Brenner a possible collaboration in sequencing the fugu genome. But this came to naught.

"That was a swashbuckling time; one in which, among other things, I learned the art of guerilla warfare in science from Sydney," Sam Aparicio observed.[10] In the summer of 2001, Aparicio attended a genome meeting at the Sanger Center. Much to his dismay—in fact his abject horror—he discovered that Lehrach, together with Jean Weissenbach of the Centre National de la Recherche Scientifique (CNRS) in France, had launched a genome effort using a different puffer fish called Tetradon. The team presented data at the meeting showing that they had already acquired about five- to six-fold coverage of the genome, whereas the Brenner group had, at best, no more than one- to two-fold coverage after a decade's worth of effort. The stunned Aparicio staggered out of the meeting and immediately called Brenner at his nearby home in Ely. Brenner suggested that he come to Ely so that they could discuss this unwelcome dilemma.

Brenner was aware of the efforts by Lehrach and Weissenbach, and knew that another genome enthusiast, Eric Lander of the Whitehead Institute at MIT, had joined that team. The critical question now was how to outpace a competent research team already well on its way—if that was, indeed, even possible. For a long while Brenner gazed abstractly out of the living room window that overlooks the picturesque garden, lovingly tended by May. Suddenly he turned to Aparicio and said with conviction: "I know how to fix this. We have to hook up with Celera and get access to their entire DNA sequencing capacity. We have to contact Craig Venter immediately and ask him whether this is possible."

Brenner contacted Venter forthwith and communicated the dilemma. Venter was a renowned maverick in his own right, and not especially fond of Lander. He, therefore, agreed to place the entire DNA sequencing capacity of Celera (somewhere between 300–400 ABI sequencers) at Brenner's disposal for the next few weeks—at cost price, an enormous sacrifice for a private company. Upon hearing this welcome news, Brenner and Aparicio determined to announce the fugu genome sequence at the 13th International Genome Sequencing and Analysis Conference in San Diego on October 26, 2001, a mere two months hence. "We calculated that if we acquired Celera's entire pipeline capacity for ten days we could do the entire thing," Aparicio related. "But this was going to cost us about two million dollars."[10] As we shall see in Chapter 25, by this time Brenner had cultivated considerable financial backing from the government of Singapore, making the effort possible. Less than a month after Aparicio learned about the rival Tetradon genome project, Brenner's group relocated the entire sequencing operation to Celera. Despite the usual maddening setbacks that seem to plague the most urgent research projects, the group completed the fugu sequence and its early annotation in time to announce the news at a press conference at the Genome Conference in San Diego. Brenner stated in the press release:

> This represents the culmination of more than a decade of work in Cambridge and Singapore. Without JGI's initiative and Singapore's strong [financial] support, the project would have languished. We already know that it will illuminate the human genome sequence and help us to understand it.[11]

In late August 2002, Aparicio, Brenner, and a long list of collaborators published the entire fugu genome sequence in the journal *Science*.[12] Soon after the paper was accepted for publication, Aparicio approached the editor of *Science* about representing the fugu genome on the cover of the issue in which it was to be published. The editor was happy to comply, provided the group could deliver an appropriate cover illustration on time. Aparicio forwarded all manner of images, none of which satisfied the editor's taste. In something of a panic, Aparicio searched the Web for suitable material, and discovered April Vollmer, a New York artist who specialized in, among other things, Japanese woodcut prints. To Aparicio's surprise and delight, Vollmer featured a beautiful puffer fish woodcut on her website. Aparicio immediately sent a copy to the editor of *Science*—who enthusiastically approved the choice. When Aparicio called Vollmer to inform her of the good news, he was careful to ask whether her woodcut was of the species Fugu. His heart sank when he learned that, in fact, Vollmer's image was of

Tetradon—the species used by the rival Lehrach group. "Of course there was absolutely no way that we could go onto the cover of *Science* with a picture of a Tetradon, because that was what our rivals were working on," Aparicio related. "That would have gone down as the biggest scientific joke of all time." He, therefore, asked Vollmer whether she would be willing to create a new woodcut—in record time. Aparicio boldly pledged that, in return for a limited run of woodcut prints, each member of the sequencing consortium would purchase one for a specified sum. "It really got down to the wire," Aparicio recalls. "I put April directly in touch with the *Science* editorial office and she had to FEDEX the final thing to them. I think she got it to them with two days to spare."[10]

Ironically, the Brenner group later discovered that the Tetradon group had pooled DNA from multiple individual fish. They were seriously delayed in completing their sequence because the many polymorphisms that existed in different species complicated their DNA fragment assembly strategies. By luck or design the Brenner consortium worked on DNA from a single fish and was spared this problem.

24

California Bound

You will always want to do more

IN THE EARLY 1980S BRENNER BECAME FORMALLY ENGAGED with the Salk Institute for Biological Sciences in La Jolla, California. Aptly described as the "jewel of Southern California,"[1] La Jolla occupies the Pacific coastal foothills just north of San Diego. The town lies along a majestic coastline, prompting a *Los Angeles Times* writer to describe La Jolla's "sequestered beauty" as a combination of "a swimmable surfer's sea, lapping limestone cliffs, palm trees mixed with pines, cool mornings, sunny afternoons, perfect air, a collegial setting, very high and quite justified salaries, and no traffic."[1]

Among the resident population of La Jolla are the students, faculty, and staff of the San Diego campus of the University of California, as well as those associated with several private research organizations, including the Salk Institute. Aside from its considerable scientific reputation, the Salk is one of the most architecturally distinguished buildings in the world. The facility was designed by the celebrated architect Louis Kahn who was instructed by founder Jonas Salk to "create a facility worthy of a visit by Picasso."[2] Perched on the edge on the magnificent Torrey Pines Mesa, the Salk features two imposing towers flanking a massive open courtyard, which reveals an unobstructed vista of the Pacific Ocean.

The Salk opened its doors in 1963 with five senior scientists (Jacob Bronowski, Melvin Cohn, Renato Dulbecco, Edwin Lennox, and Leslie Orgel). Early in its history the institute established a panel of distinguished non-resident fellows with five-year appointments, who visited on a regular basis to participate in discussions with and periodic review of the scientific staff. In the early 1980s Frederic de Hoffman, a nuclear physicist who served as President of the Salk for 18 years, recruited Brenner as a nonresident fellow with enthusiastic support from Francis Crick (who had been at the Salk since 1976). Crick wrote to Brenner:

> [E]verybody here is very keen that you should become a non-resident Fellow. The duties are not onerous and the financial reward appreciable. Moreover you can travel here in comparative luxury [meaning First Class air travel] and have pleasant "winter" holidays at no expense to yourself. Above all it would give us an opportunity to chat over things in a relaxed way, especially as it seems more difficult to do this in Cambridge.
>
> For all these reasons, and especially the latter, I do hope that you'll accept. It would give me great personal pleasure if you did.
>
> Yours ever,
> Francis[3]

Brenner's appointment was formalized in March 1981. His primary obligation was to join the other nonresident fellows (Salvador Luria, Gerald Edelman, Paul Berg, and Ed Lennox) at an annual meeting to advise on promotions, appointments, scientific proposals, and the like. Predictably, Brenner injected a spirit of rigorous review and uncensored criticism to the meetings. Equally predictably, his strong and often bluntly stated views occasionally clashed with those with more sedate dispositions. One such confrontation with Nobel Laureate Roger Guillemin, Director of the Laboratory for Neuroendocrinology at the Salk, prompted Brenner's resignation a full year before his five-year term was due to expire. This war of words arose from a consideration of the merits of a young physiologist—who shall be nameless. The actual dialogue that emerged during the meeting is, of course, confidential. But on September 5, 1985 (a good nine months after the Salk annual meeting), Guillemin wrote an angry letter to Brenner.

> Dear Brenner,
>
> At the meeting of the faculty of the Salk Institute last January in Rancho Santa Fe, you expressed not much more than contempt for the work of ——, a young physiologist in —— group whose appointment was being discussed, qualifying in your lapidary and final way as that of a mere "follower." [You stated he] was simply waiting for new peptides, characterized by others, to put through his mill of physiological testing and thus crank out reports on "pedestrian physiology," to put it simply.
>
> Why, now, this page of unfriendly prose? It is because the current August issue of *P.N.A.S.* (Vol 82, 5255) carries a report contributed by you and entitled "Identical short peptide sequences in unrelated proteins can have different conformations ... etc."

And what do you do in that paper that is different from what you castigated ——— for? Nothing that I can see. You are making use of primary structures of proteins, as established by others, for your own interest in science and drawing interesting conclusions. And of course there is nothing wrong in that. What I object to, however, is that in your case it is acceptable, while it elicited nothing but deprecatory remarks from you regarding ——— work. What is the rationale for this double standard? If you and others of the decision-making body of this place just do not consider classic physiology to be part of what the Salk Institute wants to foster, fine, let it be heard as such and that's it. I think it would be a major mistake for the future of this place; but I have done enough over the last 15 years, since I was invited to come here, to make my point clear, so as not to be interested in defending it again, particularly in the contemptuous and antagonistic relationships that you seem to favor and generate at these meetings.

There is more to biology than molecular biologists and molecular biology, in spite of the marvellous power of reductionism.

Roger Guillemin[4]

Brenner responded immediately—and emphatically.

16th September 1985

Dear Guillemin,

I will not bother to seek an apology from you for questioning my integrity because your letter is stupid as well as offensive. One of the more regrettable features of being a Nonresident Fellow of the Salk Institute is that it compels me to spend time with people like you.

Yours sincerely,
Sydney Brenner[5]

Brenner forwarded a copy of this response to Salk President de Hoffman, followed by a brief note announcing his resignation from the institution. In mid-1986, however, he served as a consultant to the president, and in 2001 was appointed Distinguished Research Professor at the Salk, where he now collaborates with neurobiologist Terry Sjenowski. He became Emeritus Research Professor in 2005.

By the late 1980s Brenner's respiratory problems had become aggravated to the extent that he suffered repeated episodes of acute pulmonary infection, especially during the inhospitable British winters. He began, therefore, to think about relocating to La Jolla permanently, or at least during the winter months. This notion was reinforced by May Brenner's health problems, which had also deteriorated in the Cambridge winters. Having helped organize a meeting called RNA Relics in the late 1980s, and with no immediate professional prospects in sight, Brenner at one time modestly considered spending his time in La Jolla organizing innovative scientific meetings while simultaneously maintaining his research program at the Addenbrooke's Hospital in Cambridge.

The colleagues and friends in California with whom Brenner shared his thoughts about his future included Richard Lerner, immunologist and founder of the La Jolla-based Scripps Research Institute (SRI). The SRI was delighted at the prospect of a formal association with the famous molecular biologist, and Lerner soon arranged for his appointment. "Being the man we all love and know so well, he would suddenly appear saying that he would be around for two months," Lerner commented. "Sometimes that translated to two days! Regardless, he was a huge scientific asset to us and was greatly revered at the institute."[6]

Brenner held the title of Scholar in Residence at SRI from 1989 to 1991 and was a Visiting Member from 1991 to 1994. Besides the welcome relief of the balmy Southern California weather, the arrangement suited him well as he was provided with laboratory space to house some of the colleagues who were with him in his final days at the Addenbrooke's. The news of his move traveled quickly across the Atlantic, prompting an administrator from the MRC to inform him in early 1991:

> Reports started reaching us a while ago from travellers through the Scripps that they are preparing a laboratory for you to occupy on a part-time basis. It would be helpful if you could let me know what you have in mind . . . ."[7]

Brenner responded:

> The matter you refer to in your letter of the 12th February 1991 seems to be of great interest to the busybodies at LMB, but it is none of your business. . . . I long ago accepted that my MRC work will be drawing to a close and it is natural for me to take steps to look after the people in my lab who are not MRC staff members and for which the Council has no responsibility.[8]

Brenner rented a (then affordable) condominium a mere block from the Pacific Ocean and conveniently close to the Scripps. Science writer Nicholas Wade once described this home: "Sydney Brenner's picture window overlooks a children's playground and a kelp-strewn beach where surfers contest the Pacific."[9] Here he could enjoy the idyllic combination of uninterrupted solitude, to indulge his ever-churning thoughts, and a haven with warm winters.

In succeeding years May, and sometimes their children and grandchildren, joined Brenner for seaside vacations in the winter months. Brenner now spends a good portion of most winters in La Jolla and was often accompanied by May when she was alive. But within a few months of her stays in California, May Brenner's patience for the intellectual poverty of the place and her general disdain for the United States usually wore thin.

During his sojourn at the SRI, Brenner collaborated with Lerner and with Kim Janda, a talented chemist at the institute. Much of their research involved encoded combinatorial chemistry, a novel technology that combines the power of genetics and chemical synthesis for screening drugs. Toward the end of the 1990s, Brenner and Janda founded a private entity called Combichem and, like many scientists who hoped to translate their scientific expertise to financial gain, they found themselves knocking on doors to acquire venture capital. Janda recalls a meeting that he and Brenner attended with a venture capital executive in the executive's office in La Jolla. Brenner was dressed in a colorful sport shirt and Birkenstock sandals.[10] In contrast the executive was attired in an expensive suit with a garish tie.

The meeting got off to a bad start when the executive, clearly aware of his own importance, began interrogating Brenner and Janda about their scientific credentials. After a few minutes Brenner, looking increasingly perturbed, reversed the interrogation. Eventually, he could contain himself no longer and lashed out at the man, asking him who the hell he thought he was. He then summarily rose from his chair and with the parting comment, "And by the way, I don't like your fucking tie either," marched out of the office. Remarkably, the company in question approved funding Brenner and Janda's enterprise. But the venture capital company's greed prompted a sell-off of Combichem long before Brenner or Janda realized any financial gains.[10]

In the midst of his engagement with SRI, Brenner entertained an enterprise with the Philip Morris Company. This giant of the American tobacco industry had made sustained overtures to the scientific community that had long been controversial. The history of these propositions dates from the early 1950s, a time when antismoking sentiments and propaganda, especially those pointing fingers at cigarette smoking as a cause of lung cancer and other respiratory diseases, reached a level that prompted reaction from the tobacco industry.

At a December 1953 meeting, tobacco heads determined to launch a pro-cigarette public relations program, which included funding research to investigate the harmful claims made about cigarette smoking. One result was an entity called the Tobacco Industry Research Committee (TIRC), created in January 1954. Through advertisements in newspapers and other media, the TIRC publicly questioned the claims of data linking cigarette smoking with lung cancer.[11] Both the TIRC and another propaganda vehicle called the Tobacco Institute were dissolved in 1998 following the Master Settlement Agreement between the tobacco industry and Attorneys General of 46 states in the United States. In the early 1970s, prior to this termination, the Tobacco Industry Research Committee spun off a granting agency, the Council for Tobacco Research-USA Incorporated, primarily supported by the Philip Morris Company.[12] A great deal of controversy surrounded the financial support offered for biomedical research by the tobacco industry. The contention among many scientists and organizations (including the American Lung Association and Americans for Nonsmokers' Rights) was that the benevolence of the Council for Tobacco Research was nothing more than a ploy to gain respectability in scientific circles. Despite these objections, a number of no-strings-attached research programs were, in fact, mounted in universities and research institutes in the United States.

In the mid-1990s Gerald Edelman,[a] then also at the SRI, was consulted by the Philip Morris Company. Charles Wang, General Council at Philip Morris, informed Edelman that the giant tobacco company was keen to improve its public image and was entertaining a significant philanthropic contribution (a sum of $250 million was mentioned). As the scientific effort remained unspecified, Edelman suggested that they consider donating money for a new research enterprise in La Jolla. Soon thereafter, Philip Morris, Lerner, and Edelman became engaged in a serious dialogue.

In June 1994 Lerner wrote a lengthy letter to Charles Wall, Vice President and Associate General Council for Philip Morris, in which he detailed plans for a scientific venture. "Please consider this letter as part of our ongoing dialogue concerning a free-standing basic research institute sponsored by Philip Morris," Lerner wrote.[13] He expressed the opinion that a leased facility would require a capital outlay of about $25 million for animal and other necessary facilities, plus an annual operating budget of $15 million. A financial package of about $250 million was eventually negotiated, with the potential of further funds in later years, a package that clearly merited serious attention.

---

[a] Gerald Edelman, a Nobel Laureate in 1972, was director of the Neurosciences Institute and president of the Neurosciences Research Foundation, a publicly supported not-for-profit organization that is the Institute's parent. He is also a professor at The Scripps Research Institute and Chairman of the Department of Neurobiology at that institution. See http://nobelprize.org/nobel_prizes/medicine/laureates/1972/edelman-bio.html.

Soon thereafter, Philip Morris publicly announced its intention to fund a new research institute in La Jolla.

At the very outset Lerner and Edelman warned Philip Morris that the general public would view their offer with considerable skepticism. The best way to forestall such a response, they suggested, was to recruit a distinguished scientist as director of the institute. The director ought to be someone with impeccable credentials who was highly respected and totally incorruptible, someone of the stature of Francis Crick or John Kendrew—or Sydney Brenner. As Sydney was then physically at the SRI and without any long-term commitments, he was an obvious choice.[6]

Brenner's ethical standards did not embrace the entrenched negative view of the tobacco industry as a source of financial support. Nor was he historically alone in this view. Duke University was largely built with tobacco funds from its founder James Duke, who also founded the American Tobacco Company. In fact, in the mid-1990s, more than half the medical schools in the United States accepted tobacco industry money, as did many of the 20 British universities in the so-called Russell Group, an elite group of universities that includes Cambridge and Oxford.[14] Directing another research organization was the last thing in the world to which Brenner then aspired. However, after some prodding he agreed to serve as a consultant and to function as acting director until a permanent head was identified. He also agreed to spearhead a search for such an individual.

In January 1995 Philip Morris released a formal proposal for the creation of The Philip Morris Institute for Molecular Sciences, "[which] would conduct basic biological research on signal transduction and related fields."[15] Plans were outlined for a new 8800 square-meter building on the Torrey Pines Mesa in La Jolla, a majestic piece of prime real estate and home to the Salk Institute, the campus of the University of California at San Diego (UCSD), the Scripps Research Institute, the La Jolla Cancer Institute, and a host of biotechnology companies.

At no time did the tobacco company explicitly or implicitly suggest that the research program would in any way focus on the effects of tobacco on human health. "We have a chance to contribute to something that's truly unique and has the potential to make a highly positive contribution to our society at a time when funds for these types of enterprises are not generally available," stated a spokesperson for Philip Morris.[16] In return the company asked for and received an agreement on licensing rights of inventions at the institute, "but only if they have no application to its tobacco business."[16]

Brenner knew full well that he would be the object of considerable criticism from the scientific community for accepting tobacco money. Indeed, several nonprofit organizations dedicated to human health, including the American Lung Association and Americans for Nonsmokers' Rights, vigorously opposed the plan. However, Brenner was undeterred by what he viewed as a sort of self-righteous posturing. "Most financial endowments come with a tainted history of some sort and I was of the firm view that in due course history would dignify Philip Morris's efforts."

Brenner appointed a blue ribbon Board of Scientific Directors to screen applications from outstanding young investigators fresh out of their post-doctoral training. These hand-picked lucky few would be appointed to the staff of the new institute for a maximum of five years; during this time they would be expected to conduct independent, innovative, high-risk research. Brenner planned to serve as an all-purpose scientific mentor—a sort of provost. "In the final analysis I believed that if this enterprise launched the career of just one brilliant young investigator it would be worth the effort and cost."

This plan represents the first of several examples of Brenner's commitment in the later years of his career to identify and promote highly promising young scientists. In so doing he has never fully abandoned the pursuit of his own scientific interests and, at the time of this writing, actively oversees modestly sized research groups in several parts of the world. The closure of the MRC Unit for Molecular Genetics at the Addenbrooke's Hospital, however, marked the end of Brenner's desire to mount a substantial research enterprise of his own. "When you reach my age the long-term and the short-term become the same thing," he often says. Henceforth, he wished primarily to leave a legacy of truly outstanding young scientists who directly benefited from his vast experience and from his scientific vision for biomedical research in the future.

Efforts to identify a scientific director for the Philip Morris Institute for Molecular Sciences floundered. "I spoke to a lot of people, but no one wanted to touch it. They didn't want to be involved with tobacco money." Eventually, a suitable candidate was identified and was actually house hunting in La Jolla when Brenner was asked to visit the Philip Morris head office in New York. There he was summarily informed that the enterprise was being shelved for the foreseeable future—business talk for shelving the project permanently. In view of the enormous amount of time and energy he had devoted, in the same breath the company offered Brenner $2 million a year for a period of five years. The idea was for Brenner to mount a new research initiative of his choice—wherever he could find appropriate space.

Armed with this unanticipated financial largesse, Brenner identified rental space in a facility owned by the University of California at San Diego, where he launched a new entity called the Molecular Sciences Institute (MSI) to continue his genome research and to mount a program in bioinformatics. However, the university apologetically informed Brenner that it had discovered an urgent need for the space he occupied. Determined to find other alternatives, Brenner visited with the leadership at the nearby Salk Institute, offering dollars from his Philip Morris money in exchange for laboratory space. But the Salk leadership, too, was concerned about negative connotations associated with *any* alignment with the tobacco industry and declined the offer. Once again Brenner found himself in the peculiar position of having $10 million to spend—and no place to spend it.

Brenner's many admirers then included a young Harvard assistant professor, Roger Brent. A native of Mississippi, Brent had earned a bachelor's degree in computer science and mathematics from the University of Southern Mississippi and a doctorate in biochemistry and molecular biology under Mark Ptashne's mentorship at Harvard. Brent had long admired Brenner and had corresponded with him since the mid-1990s, sharing his passion for genomics and his escalating interest in applying bioinformatics to biological systems. Reciprocally, Brenner identified in Brent an agile young mind with an obvious interest in genomics. More important perhaps, Brenner recognized that Brent provided an opportunity to address another stumbling block to establishing MSI—the need for a substantial infusion of nontaxable state or federal grant support in order to qualify for nonprofit status in the state of California. "There was absolutely no way that I was going to apply for grant support at that stage of my life," Brenner stated. He, therefore, began to consider persuading an established younger American collaborator to join MSI—someone in particular who might bring grant support to the place. Enter Roger Brent.

When contacted by Brenner and informed about the plan, Brent was excited. But having spent his brief scientific career in academia, he was not keen to forgo its many advantages. However, Brenner urged him to consider the problems associated with mounting and sustaining a bioinformatics-intensive research program in a university. Arguing that the multidisciplinary relationships required for such an enterprise would be daunting for an academic institution, Brenner pointed out especially the many competing agendas that routinely confront these organizations. Brent was convinced of the validity of Brenner's arguments. Besides, he was optimistic about the future of bioinformatics and was confident that he could obtain the necessary funds from the private sector and from traditional granting agencies. Nonetheless, Brent recognized that it would be foolhardy to completely sever his academic ties at such an early stage of his career. Therefore, he insisted that his involvement with the institute include at least a formal appointment in a nearby academic institution, "even though Sydney viewed that kind of security as the small fig leaf that it truly was."[17] It was clear that university affiliation importantly offered the ability to seek grant support. The University of California at San Francisco agreed to appoint Brent as an adjunct professor.

Once he had identified a suitable facility on Shattuck Avenue, Berkeley (close to the bustling campus of the University of California) architect Ken Kornberg (son of the late biochemist Arthur Kornberg and brother of scientist Roger Kornberg) was engaged to design the new laboratories. By 1998 MSI was up and running. Having come to the decision that La Jolla was a likely long-term winter refuge, Brenner purchased the condominium he had rented there and commuted to Berkeley once or twice a month.

It is probably fair to state that Brenner never possessed the level of commitment to MSI that Brent did. At this stage of his life, Brenner regarded the institute as but one of several ongoing scientific adventures. Brent, on the other hand, viewed MSI with more serious long-term goals. He was determined to turn MSI into an entity with a secure future, but he perceived that Brenner was less concerned about obtaining sufficient funds to make this a long-term proposition. "These sorts of issues occupied a limited extent of his mental space," Brent explained. "He would pop up from La Jolla once or at most twice a month, sometimes once every two months, and by 2001 his lack of evident seriousness about stabilizing and building the institute lent the enterprise a contingent quality that left me feeling less than sanguine."[17] Indeed, in that year Brenner ended his formal association with MSI, and Brent became Director and President. Brent retains enormous respect and admiration for Brenner's scientific capacity. "He stands to molecular biology as Thomas Jefferson stands in relation to the United States of America," he once told an interviewer.[18]

Then approaching the age of 75 and no longer tied to any formal scientific enterprise, Brenner might have been expected to consider pursuits requiring less effort—organizing innovative scientific meetings, consulting, writing books, and pursuing his eclectic reading tastes. Nothing could be further from the truth. As for his diminishing health, "Health problems are to Sydney nothing more than interferences to be dealt with," said Brenner's friend and colleague Philip Goelet.[19]

Brenner's visits to the Berkeley area prompted a renewed acquaintance with Sam Eletr, a physicist with a long-standing interest in designing, building, and marketing machines for sequencing (and otherwise analyzing)

DNA. Eletr first encountered Brenner in the early 1980s when he visited the Rothschild group in search of venture capital to establish the company called Applied Biosystems, which became hugely successful. The entrepreneurial physicist-turned-biotech developer impressed Brenner, whose favorable recommendation to Rothschild helped Eletr secure the venture capital he sought. Applied Biosystems was sold to Perkin-Elmer in 1993.[20]

Brenner and Eletr lost contact until the early 1990s when Eletr was busy launching another company, Lynx Technologies, and offered Brenner a place on its scientific advisory board. At that time Brenner was toying with ideas about a technology that eventually became known as massive parallel DNA sequencing.[20] Indeed, Brenner was lead author on two papers that addressed the feasibility of this technology.[b,c] Business decisions, including a merger with a company called Solexa, Incorporated (Lynx and Solexa were eventually acquired by an entity called Illumina) throttled this venture. However, Eletr continued to enjoy extensive interactions with Brenner and benefited greatly from his advice and council. "Whenever I posed a technical 'what if' question to Sydney he'd find a way of answering it," he explained.[20] One of the "what ifs" concerned the adaptation of his multiplex DNA sequencing technology to analyzing DNA from multiple different individuals. This idea led to the establishment of yet another private research entity, Population Genetics Incorporated, in Cambridge, conveniently close to Brenner's home in Ely.

Brenner and Eletr, together with Philip Goelet, developed Population Genetics with the intent of bringing technologies to the market place for interrogating multiple genomes simultaneously. By using comparative genomics and techniques for rapidly screening human populations, they hoped to identify rare polymorphisms and disease signatures. Perhaps no one has articulated the urgent need for more rapid methods for DNA sequencing than Brenner himself. "Sequencing will always be rate-limiting,

---

[b] Massive parallel DNA sequencing involved an open-ended platform that analyzes the level of expression of virtually all genes in a sample by counting the number of individual mRNA molecules produced from different genes. See Reinartz J, et al. 2002. Massive parallel signature sequencing (MPSS) as a tool for in-depth quantitative gene expression profiling in all organisms. *Briefings in Functional Genomics and Proteomics* 1: 95–104.

[c] Brenner S, Johnson M, Bridgham J, Golda G, Lloyd DH, Johnson D, Luo S, McCurdy S, Foy M, Ewan M, et al. 2000. Gene expression analysis by massively parallel signature sequencing (MPSS) on microbead arrays. *Nat Biotechnol* 18: 630–634; Brenner S, Williams SR, Vermaas EH, Storck T, Moon, K, McCollum C, Mao JI, Luo S, Kirchner JJ, Eletr S, et al. 2000. In vitro cloning of complex mixtures of DNA on microbeads: physical separation of differentially expressed cDNAs. *Proc Natl Acad Sci USA* 97: 1665–1670.

because you will always want to do more," he told an interviewer. "... The reality is that you don't want just one hemophilus genome; you want to do the genomes of dozens of hemophilus strains."[21] Convinced of the imperative for rapid DNA sequencing, Brenner is confident that new technologies will appear on the scene. "These will involve not just incremental improvements of what we have seen. Instead one will be looking for techniques that give three to four orders of magnitude improvement."[21]

Nowadays Brenner roams the world proselytizing about unlocking further secrets by using comparative genomics: "With the right technology it is biologically and mathematically possible to extract DNA from every human being alive today and examine these in solution," he told an enthralled audience in Pretoria, South Africa, during one of his recent visits to the country of his birth. "We could have the entire global human genome in a few micrograms of DNA. Just imagine that! What remains to be done is to develop the technology—no small thing. But I think it can be done."

Brenner is decidedly critical of bioinformatics strategies that rely on knowledge bases of molecular biology (such as the Gene Ontology [GO] project) to unlock the secrets of evolution. "I suspect that the best that gene ontology will do is give us a common language in which to express our confusion," he wrote in a typically Brennerian piece entitled "Ontology Recapitulates Philology."[22]

> My aim is to get out of the Tower of Babel and go somewhere else, rather than try to find a common language to govern it. The connection between Babel and babble is more than a coincidence. Meeting someone with a name tag that says "I'M CHUCK," tells me nothing about the immense biological object carrying it and it might just as well have said "MY NAME IS CHUCK" and, perhaps in smaller print, "and who I am is my business!"

In a lecture entitled Genes, Genomes, and Evolution delivered at Columbia University in celebration of the 50th anniversary of the double helix, Brenner used another analogy. "Think of the genome as the white pages of the telephone directory," he told his audience. "What people are thinking about doing now is to compile the yellow pages; this is the annotated genome. It's a great thing to know there are seven plumbers on one block. But you will still have a fragmentary description of genes and gene function in a cell."[23] Brenner doesn't believe that thinking of genes that way is the correct level of abstraction. He prefers to think of gene function in the context of cell function, and since "all projects should get a name" he calls this notion and its study, the CellMap project.

The rational for the CellMap project is more fully explained in a remarkable paper that Brenner wrote on the occasion of the 350th anniversary of the Royal Society in 2010. Brenner entitled this article Sequences and Consequences.[24] The article is an unrestrained attack on the discipline called systems biology. Decrying the vast accumulation of sequence information, protein–protein interactions, and other massive data-generating "omics" efforts, Brenner pleads for a theoretical basis to unify biology. "It is only theory that will allow us to convert data to knowledge," he lamented. "Molecules tell us nothing about cells and their behaviour, and neurons tell us nothing about brains and how they work. We must avoid using observations to deduce models." The claim that systems biology "can solve the inverse problem of physiology by deriving models of how systems work from observations of their behaviour" is wrong. Brenner's arguments are reminiscent of the Quine–Duhem thesis, which argues that "it's impossible to test a single hypothesis on its own, since each comes as part of an environment of theories. Thus we can only say that the whole package of relevant theories has been collectively falsified, but cannot conclusively say which element of the package must be replaced."[25]

Scientifically-oriented business individuals like Eletr and Goelet are a potent source of stimulation to Brenner. Goelet comments admiringly on Brenner's willingness to board a plane and travel halfway round the world to promote ideas, and if unsuccessful, to quickly get on another plane and go in the other direction to find the support he seeks. Goelet believes that scientists of Brenner's stature are not really interested in money to line their pocket books; they want money to support their research. In contrast, business types are tied to a bottom-line orientation and are not interested in science with little financial potential.

> Sydney doesn't ask himself questions such as: "What are the big unmet needs in the life sciences market?" Rather his focus is: "How do I get investors to give me money to solve a scientific problem that I'm interested in?" In that regard he sees himself as a Michelangelo in search of a wealthy patron to support him.[19]

In these endeavors Brenner's primary role is to sell the science, while Goelet's is to make sure that the project is sustainable by attending to the business end.

As for Brenner's opinion of the business world, "Dealing with the private sector, especially venture capital types, is an interesting experience," he once stated. "I learned how the VC folks combine stupidity with greed—and how the pharmaceutical companies combine stupidity with stupidity." Though perhaps not in the same breath, he has also said, "DNA makes RNA makes protein—makes money."[26] Consider the grim reality that about 90% of venture capital investments fail commercially. Some have heard Brenner indicate that he would be delighted if a venture capital type judged one of his start-up schemes as likely one of the 90% failures—and simply gave him the money!

25

Singapore

*I was interested in helping a country motivated to
change in the right direction*

A BRITISH CROWN COLONY SINCE 1867, SINGAPORE ENJOYED relative affluence as a major British port of call for ships traveling between East Asia and Europe. The growth and prosperity that endured until the early years of World War II was interrupted by the Japanese invasion on December 8, 1941 (a day after the attack on Pearl Harbor). The tiny country endured harrowing times until the British liberated it in 1945. Following the end of the war, Singapore became increasingly independent of Britain. In 1963 Singapore merged with the Federation of Malaya, but the merger was short-lived. Ex-Prime Minister Lee Kuan Yew, one of the heroes of Singapore's modern economic ascendancy wrote: "Within a year, in July 1964 ... we were trapped in an intractable struggle with Malay extremists of the ruling party ... who were intent on a Malay-dominated Malaysia. ... By August 1965 we were given no choice but to leave."[1] On August 9 of that year, Singapore became an independent republic.

With the advantages of a small but resourceful and forward-looking political leadership, Singapore created a modern and successful economy. English was declared the official language of the mixed Chinese, Indian, and Malay population. An Economic Development Board was empowered to orchestrate the emerging manufacturing and service sectors and instituted tax incentives to attract foreign investment. The mid-1980s witnessed major technological developments, including a nationwide broadband network before the end of the 20th century. Now visitors entering Singapore's ultramodern Changi Airport are struck with dazzling images of growth, opulence, and impeccable organization. The center of this immaculate city-state (with a population of 4.5 million people) teems with skyscrapers, plush restaurants, and high-end boutiques. Its wide boulevards are crowded with alert and confident-looking teenagers of Chinese, Indian, and Malay descent.

For many years after it became independent, Singapore enjoyed the advisory and consultative services of Victor Rothschild. His motives were not entirely altruistic, of course; Rothschild was but one of many foreign entrepreneurs seeking to profit from Singapore's burgeoning economy. Recall from Chapter 22 that Rothschild had solicited Brenner as a consultant to his biotechnology enterprise. As early as 1981 he encouraged Brenner to visit the tiny Southeast Asian country to examine its biomedical research and technology infrastructure. In September of that year, he wrote to Brenner:

> I have just received a letter from the Deputy Prime Minister of Singapore saying that they are not really ready for a molecular biology VIP such as yourself.[2]

But a few years later the Singapore government was eager to meet Brenner and to hear his views on biotechnology development. Brenner visited the country in September 1983, to meet with Prime Minister Lee Kuan Yew and Deputy Prime Minister Goh Keng Swee. Swee, a graduate of the London School of Economics, is the man most frequently credited with bringing about Singapore's economic miracle. Brenner also presented a lecture entitled Overview of Biotechnology in Industry[3] in which he stressed the importance of entrepreneurial relationships among science, technology, and industry. However, he cautioned that less-developed countries, such as Singapore, had enormous barriers to cross in biotechnology, an enterprise critically based on high-level science and a strong science infrastructure. "The way to succeed in biotechnology," he cautioned "is to start at a level where you can assimilate the technology and make it the basis for future scientific investment. But," he concluded, "while concentrating on the practical questions on how to apply biotechnology here, you must not forget about the need for a scientific platform on which it is to be based."[3] Essentially, Brenner's message was, you cannot successfully compete in the biotechnology sector unless you first develop a robust infrastructure in modern biomedical research.

Some of the more fiscally conservative elements in the government expressed ambivalence about mounting such a grand enterprise, especially one that might not pay dividends in the future. "We are still primarily a country of technicians," they plaintively told Brenner. "And you'll remain a country of technicians unless you get seriously involved in cutting-edge research and development," Brenner responded. This comment resonated with the political leadership of the country. Brenner's reputation as a scientist was well known. More important, he impressed his hosts with his honesty and his confidence about confronting complex issues.

During a later, more extended visit, Brenner again met with Yew, Swee, and other leaders in the Singapore government—those who were interested in mounting biotechnology in the country. In later years when questioned as to why he was spending so much time in Singapore, he responded,

> I was genuinely interested in helping a young country motivated to go in the right direction. ... This was to be an experiment in developing state-of-the-art biomedical research at a national level in what was a third-world country not too many years before. I viewed it as an exciting venture and an exciting opportunity.

Brenner was also enthusiastic about the potential for placing his own brand on the culture of modern molecular and cellular biology in a place like Singapore. Although lagging in some areas of technology development, it was most certainly not short of financial resources. Finally, and perhaps most important, Brenner was impressed by the enthusiasm and determination he witnessed for developing cutting-edge biomedical research and biotechnology in Singapore. Brenner related:

> They asked me to give them a written proposal. Knowing that busy people don't like reading lengthy documents and mindful of Winston Churchill's famous admonition that he didn't like reading anything that was more than one side of a single sheet of paper, I wrote out a basic plan for the future on a half of one side of a single sheet of A4!

The central features of this proposal were quickly endorsed.

Before leaving the country Brenner was introduced to the owner of a flourishing Singaporean mushroom company and invited to join its Scientific Advisory Board. Welcoming the invitation, Brenner requested that he be compensated in mushrooms—preferably of the shitake variety. When he returned to the United Kingdom, he did so with mushrooms in hand. "I am writing to thank you for the mushrooms which arrived with me in very good condition and which I and some friends enjoyed enormously," he penned his Singaporean colleague. "I hope that our future meetings will have as delicious an outcome as the last one."[4]

Once committed, the Singapore government was not hesitant in moving ahead. A new building was erected on the campus of the Singapore National University for housing an Institute for Molecular and Cell Biology (IMCB) that would serve as the nucleus for mounting a molecular biology presence. When the directorship of the institute was predictably offered to Brenner, he suggested that an established, well-respected Singaporean scientist working abroad be persuaded to return to Singapore to lead the new venture (this was to be someone of their choosing). However, he offered to assemble a Scientific Advisory Board to oversee operations at the IMCB and agreed to serve as its first chairman. Three names surfaced as candidates for director of the new institute. Chris Tan, an immunologist in Canada was making impressive progress with his research on interferon. Nam-Hia Chua was an established and well-regarded molecular biologist at the Rockefeller Institute in New York, and Louis Lim was then a respected neurobiologist at the National Institute for Neurology in London.[5]

Tan received a personal phone call from Deputy Prime Minister Goh Keng Swee, imploring him to return to Singapore to head the planned IMCB. Urged to develop a formal proposal that could soon be shared with Brenner, Tan completed the proposal in two weeks and sent it to Singapore by diplomatic pouch. Tan then flew to London to meet with Brenner. In short order Tan was off to Singapore to take up his new position, but without relinquishing his faculty appointment at the University of Calgary, to which he intended to return at some point.

The new institute, modeled after the Whitehead Institute at MIT, was erected on the doorstep of the medical school of the National University of Singapore. Constructed in a mere thirteen months, this was the first of multiple building projects completed at such breakneck speed. The ICMB officially opened its doors on October 2, 1987, with 38 research scientists and Tan as Director. The facility included a laboratory for Brenner to continue some of his fugu genome research program, relocated from Cambridge, under the supervision of Byrappa Venkatesh, who moved to Singapore permanently.

When he officially opened the new institute, Minister of Education Tony Tan Keng Yam stressed the government's commitment to the realization of three important objectives. The goals were to identify and train Singaporean intellects to develop biotechnology in the country, to provide the infrastructure required to support biotechnology in Singapore, and to attract leading foreign scientists to the country to conduct innovative basic research that might lead to practical applications. Chris Tan oversaw an increase in the staff of the ICMB from 35 scientists in 1987 to more than

500 by 2003. Having succeeded admirably launching the new institute, Tan returned to Canada in 2004.

Not everyone in Singaporean government circles, however, was pleased with the association between the IMCB and the National University of Singapore. To ensure that the new institute was not mired in traditional university bureaucracy, an extraordinarily competent and dynamic member of the Singapore leadership—Philip Yeo Liat Kok—became increasingly interested in and committed to biotechnology and installed himself in the epicenter of the planning effort. Yeo, an industrial engineer with a Harvard M.B.A., served in various capacities in the Singapore Ministry of Defense before assuming the chairmanship of the country's Economic Development Board in January 1986. In the newly constructed science complex called Biopolis, a profusion of documents and architectural blueprints (in seeming disarray) covered Yeo's massive desk. Without effort, he was able to dive into this morass of paper to retrieve a relevant document at a moment's notice.

Yeo's absolute dedication to developing cutting-edge biomedicine and biotechnology in Singapore and the energy he brings to that goal are truly impressive. The extent of his faith in Brenner, with whom he quickly identified, cannot be underestimated. "The key for Singapore is developing human capital, because the country is so small and resource-scarce," Yeo told an interviewer in 2001.[6] He preaches this mantra consistently. In the midst of persuading David Lane, a distinguished British scientist, to head the challenging effort that he and Brenner were orchestrating, Yeo invited his guest to a working dinner at a restaurant at the top of one of Singapore's many skyscrapers. While discussing Singapore's proud history and describing the dearth of natural resources in the tiny country, he pointed to a window and said, "David, look out of the window and tell me what you see down there." "A lot of people," was Lane's somewhat perplexed response. "Exactly," Yeo responded. "That's all we really have in Singapore—people. They are our most precious resource!"[7]

Prior to Yeo's alliance with Brenner, the oversight of matters such as the development of biomedicine and biotechnology was the province of a Science Council composed largely of senior university officials. Distinctly skeptical of academia in matters of business, Yeo was dissatisfied with the slow pace and limited vision of this administrative body. "In my experience university professors are great thinkers—but they are terrible executors!"[8] he proclaimed in rapid-fire English. Yeo then orchestrated a new entity with broader powers, the National Science and Technology Board (NSTB), and persuaded Brenner to serve on the board.[8]

Neither Yeo nor Brenner was impressed by how the NSTB operated. One particular concern was the board's intention to establish another new institute at the university to promote agrobiology, an initiative they deemed uninspired and potentially divisive. Brenner still chaired the Scientific Advisory Board of the IMCB and considered the NSTB poorly focused and ineffectual.

> These people were essentially a bunch of administrators primarily interested in making money, with little appreciation for academic research and development. I stressed to both Philip and the NSTB that the only way that biomedical research and commercial spin-offs from such research could evolve and prosper in Singapore was by establishing a single agency that oversaw *all* biomedical research, from molecules to mind—and nothing else. I told them that countries in the world that had made substantial contributions to biomedical research, such as the US, the UK and France, all had national organizations specifically dedicated to this goal; the NIH in the US, the MRC and the Wellcome Trust in England and the CNRS in France.

The Singapore government, prompted by Yeo, heeded Brenner's advice and established another administrative entity, the Agency for Science, Technology, and Research (A*STAR) that would, henceforth, promote, fund, and oversee *all* biomedical research in Singapore. Impressed once again by how Yeo made and executed decisions, Brenner introduced him to various leaders at the MRC and the Wellcome Trust in England to observe how these entities functioned. Resigning from his position with the Economic Development Board, Yeo assumed control of the A*STAR, and then made another far-reaching decision.

In Yeo's view, the broad oversight of biomedical research in Singapore required a new facility that was physically—and programmatically—independent of the university. Construction of the 2 million square feet complex of glass and chrome buildings to be called Biopolis, named by Brenner, was to begin immediately. Upon its completion the IMCB would move to Biopolis and its former site, recently erected on the university campus, would be put to other uses. A 2003 editorial in *Nature Medicine* described the new facility: "Biopolis is a concrete and glass demonstration of the Singapore government's determination to establish the city-state as the premier biosciences center in Asia."[9]

Yeo's new plan exceeded even Brenner's expectations. Yeo had selected the architects and building contractors himself and launched construction on December 6, 2001, insisting that building construction proceed around the clock, seven days a week. In October 2003 the first building—which now houses the IMCB—was completed, and an additional six buildings were well on the way.[8] By 2005 Yeo's agency had spent US $300 million on Biopolis, to create research facilities dedicated to genomics, bioinformatics, bioprocessing, and bioengineering. *Nature*, which regularly communicates progress in biomedical research in Asia in general and Singapore in particular, described the Biopolis effort in 2005:

> It is part of a multibillion-dollar investment in biomedicine designed to nurture industries in biotechnology and drugs. Yeo [and Brenner have] led from the front, enforcing an aggressive schedule and attracting some of the biggest names in biology to head the complex's lavishly equipped labs. "Yeo has so much energy. If you tapped it you could light a small city," says cell biologist Axel Ullrich, who was recruited from the Max Planck Institute in Germany.[10]

Chris Tan was succeeded by David Lane, widely known for his seminal contributions to our understanding of the *p53* gene. Then working at the University of Dundee, Lane first met Yeo in 1999 at a party in London. In search of investors for a biotech company, Lane quickly recognized that Yeo was intrigued by high profile scientists who were fundamentally interested in the commercial potential of biomedical research. "He literally knocked me off my feet. He was just so dynamic," Lane recounted.[7] Lane soon received a telephone call from Brenner, asking him to take over as chair of the Scientific Advisory Board of the ICMB, Brenner's position from which he was shortly retiring. Impressed with Yeo and Brenner's grand visions, which they translated into reality, Lane agreed to assume the role, which required that he visit the ICMB three times a year. Lane enjoyed Singapore and, in due course, accepted an offer to succeed Chris Tan as full-time director of the institute. However, Brenner had bigger plans for the capable Scottish scientist. "One fine day he dropped in on me and casually stated: 'David, you're going to take over as chairman of the Biomedical Research Council.' "[7] (The council is an administrative body that oversees the entire Biopolis enterprise and administers an extramural research program that includes the university.)

With Brenner's encouragement, Lane also initiated a program in experimental therapeutics as part of a larger translational research effort. The university facility occupied by the IMCB before its move to Biopolis now houses and trains physician scientists interested in all aspects of translational medicine. Here Lane promotes Brenner's ambition of mounting "bedside to bench" research. Under this umbrella Brenner searches out families with what he calls extreme phenotypes—rare human disease states that may be associated with single gene mutations. Lane and his colleagues recently identified a family in which all members are unable to experience the sensation of pain. "Consider the interesting perspectives that finding the relevant gene may hold for understanding the biology of pain," Lane said with obvious excitement.[7] He is persuaded by Brenner's conviction that marrying modern human genetics with modern high-throughput sequencing techniques will set the stage for abandoning animal models and studying humans directly as experimental subjects. "As always with Sydney, one thinks long and hard about what he says—because he's so visionary—and so often right."[7]

Lane is sympathetic to the Singaporean goal of leveraging biotechnology from its investment in biomedical research. In his view all academics should be fundamentally engaged in some enterprise of practical importance. "If one is a university professor one should be engaged in teaching," he stated. "The way I see it, if one works in a research institute one should be fundamentally engaged in something related, like deriving commercial benefit from one's research efforts."[7] These ideas are particularly apt in a country like Singapore that aspires to promote economic growth by generating new jobs and new markets.

Others have since followed David Lane to become part of the ambitious enterprise—Edison Liu (a prominent U.S. cancer biologist) and Axel Ullrich (former Director of Molecular Biology at the Max Planck Institute in Martiensried, Germany), and Nancy Jenkins and Neal Copeland. These two leading American mouse geneticists spent two decades at the National Cancer Institute in Bethesda, Maryland, before yielding to the irrefutable logic that when the NIH suffers a funding crisis (which has occurred with alarming frequency in recent years), their extensive mouse program is threatened. Copeland and Jenkins had been frequent visitors to Asia for well over 20 years, and the couple had become deeply attached to Asia—especially to Singapore. When Copeland and Jenkins were offered space and resources to maintain their extensive mouse colony—with the proposal that Copeland become director of the IMCB—the pair decided to take the opportunity.[11] Joining soon thereafter were Alan Colman, who advanced the field of nuclear transfer with the cloned sheep Dolly; Jackie Ying from MIT, a renowned chemical engineer; Yoshiaki Ito from Japan, known for his studies of cell transformation and tumor suppression; and developmental biologist Roger Pederson from Cambridge, who heads a new stem cell research program.

When Brenner visits Singapore, which he does regularly, he rarely sits back contentedly, indulging in self-congratulation. On the contrary, he is always restlessly seeking new and better projects, new and better technologies, new and better ideas. David Lane has observed:

> One comes to realize that Sydney really doesn't enjoy life when things are too settled. He believes there's a positive energy that flows out of a certain degree of instability and chaos. Those who want things to be very structured and stable don't suit Sydney's personality very well. He's the sort of guy who enjoys tossing the hand-grenade around. He asks the tough questions and if he sees things looking too settled and not moving forward in new directions he stirs things up.[7]

Brenner has, on occasion, considered the idea of making Singapore his permanent home. However, he is not especially keen on the equatorial weather and finds the constancy of the temperature somewhat stultifying. Mainly, one suspects that permanence is something of an anathema to Sydney Brenner; that he is, in fact, comfortable in his peripatetic life style, unlikely to call anywhere his permanent home. Like his place in La Jolla, and Philip Goelet's spacious residence in the Maryland countryside, Singapore has become another comfortable home away from home—and that suits his needs well enough. Singapore provides a welcome break from an otherwise hectic life style, so that he can relax and think science and devote time to other intellectual pursuits.

When visiting Singapore Brenner enjoys the hospitality of his colleague and friend, former student Byrappa Venkatesh, who manages Brenner's laboratory in the IMCB. There he is treated as a member of the Venkatesh family, visiting their home whenever he is in the country for more than a few days.[12] Brenner and Venkatesh often spend time together outside the lab perusing books at the local Borders bookstore. Sometimes Brenner purchases a book or two; these may feature any subject from mathematics to Chinese cooking to Tibetan history to Japanese swords. Sometimes, after several hours of happy browsing, Brenner buys nothing at all.[12]

When diagnosed with colon cancer in the early 2000s, Brenner elected to be treated in Singapore. At the time of this writing, he shows no signs of residual disease. To the dismay of the physicians monitoring his post-operative course, Brenner resumed his hectic travel during treatment with chemotherapy and anticoagulants. While walking up a flight of stairs in Switzerland, he fell and sustained a relatively mild injury. However, the ensuing bleeding became difficult to control because of his anticoagulated state and he required immediate hospitalization. Once stabilized he was rushed back to Singapore. "I was shocked when I saw him," Venkatesh remarked. "He had a ton of bruises. But he made a remarkable recovery."[12] Predictably, Brenner was soon once again flitting around the world.

Yeo never loses sight of the critical importance of promoting the advancement of native Singaporians. Among the many problems that challenged him when he assumed direction of A*STAR, few aggravated him more than the shortage of qualified local biomedical scientists. In 2001, therefore, he launched one of the most ambitious educational programs not only in Singapore, but anywhere else in the world—a program designed to cultivate a new generation of outstanding investigators.[8]

Relentless in his pursuit of financial resources for this grand experiment, Yeo generated funds—sometimes at the expense of other projects that he deemed less urgent—to train the best and brightest students at prestigious overseas colleges and universities. Participants in this program are selected from the 12,000 students who graduate each year from Singapore's high schools, some comparable to the most elite private schools in the world. Those with perfect (or nearly perfect) scores are invited to apply for financial support from the government to seek admission to a distinguished foreign university of their choice. Students may major in any discipline, except economics and law. "We have no shortage of lawyers and economists in Singapore!" Yeo explained.[8] The program, which supports travel, full tuition and board, as well as a generous stipend, maintains demanding standards. If any foreign university selected is good enough for Singapore's best and brightest, the best and brightest must be good enough for that institution. If a foreign institution declines an application from a Singapore high school senior, the student is dropped from the program. Once admitted, and while studying towards a bachelor degree, students are expected to maintain impeccable academic records to remain in good standing in the program.

Following graduation these scholars must return to Singapore to spend a year in a biomedical research laboratory, where they begin their training as future scientists. Then follows another grueling challenge; each student must seek admission to an overseas graduate program leading to a Ph.D. or equivalent degree. Once again, only the most prestigious universities are considered appropriate, and successful applicants receive a complete financial assistance package from Singapore. With doctoral level degrees from universities such as Stanford, Harvard, MIT, and Cambridge, students are bonded to return to Singapore. For at least five years, they must engage in postdoctoral work and then seek junior faculty positions in the Biopolis consortium—or elsewhere in the country. When asked what would happen to a student who might balk at this imposed five-year bond, Yeo's reply was unequivocal. "They or their family then owe me a lot of money."[8]

In its fifth year (at the time of this writing), the program supports about 700 students at various levels of training. Yeo anticipates that by 2010 he will have given out about 1000 scholarships. "If only 50% of these individuals decide to remain in Singapore I'll be happy," he states. "That's a pretty good return on an investment."[8] If only a small percentage of these ultimately decide to engage in Singapore's biotechnology sector, Yeo will also be satisfied. "The buildings you see around you are all well and good," he stated. "But the scholarship program is my real investment in Singapore. By 2020 these individuals will operate the entire biomedical research and technology enterprise in Singapore."[8] Yeo maintains a comprehensive database of all students in the program and routinely monitors their academic progress.

"There are problems and issues here like there are everywhere," David Lane commented. "But in the final analysis the question I ask myself is: 'Can I achieve things here?' And the answer so far is: 'Yes, I can.' In fact that's the most amazing thing about Singapore—how much one can achieve here."[7]

In part, due to his achievements in the development of a science infrastructure and biomedical research in Singapore, Brenner has attained iconic status in Singapore and is an honorary citizen of the country. The VIP section of the stunning National Orchid Garden, arguably the largest and most comprehensive orchid collection in the world, includes a hybrid named for him. The campus of the National University of Singapore houses the Brenner Center for Molecular Medicine, a facility dedicated to translating the basic sciences at Biopolis to improving health and patient care; and the library of the Institute for Molecular and Cell Biology at Biopolis also bears his name. In late 2006 Brenner was awarded the Singapore National Science and Technology Medal. When formally accepting this honor the normally taciturn Brenner stated:

> This award means much more to me than all the others I have because here in Singapore I can see the tangible results of it. Here we now have hundreds, no, thousands of young people devoted to science and to a career in biomedical research—and that's an opening to the new world. I am very proud of what has been accomplished here.[13]

Brenner is effusive in his praise for Yeo. In his acceptance speech he stated:

> A lot of my work here is owed to my close collaboration and friendship with Mr. Philip Yeo. This collaboration has been one of the most exciting

things I've participated in. In the video [shown on this occasion] I said we went to a place where you now see Biopolis and there was nothing there. We went back a couple of weeks later and everything was cleared. A couple of months later buildings were rising, a few months further on people were working there. Biopolis appeared almost overnight. Colleagues of mine have said that if this were to be attempted in their country it would take 10 or 12 years to accomplish. This is one of the great things I have come to realize about Singapore; that they are impatient and everything must be done yesterday—not today, but yesterday. It is very stimulating—and it actually produces results.[13]

26

Mentoring Again

The best way to influence science is to give high-level advice

Aorganizations sought his advice and forward-looking vision. This chapter
relates two other enterprises that Brenner identified as fruitful opportunities
for promoting cutting-edge research; one in Okinawa, Japan, the other in the
United States, in Virginia, where the Howard Hughes Medical Institute was
planning a major paradigm shift.

> A remote island best known for its U.S. military base, scuba diving and
> goya (an intensely bitter melon) might not seem the obvious place to build
> a cutting-edge research institute from scratch.... But the Japanese govern-
> ment has been pouring cash—Y3.3 billion (US$30 million) over the past
> two years—into a new institute to be based in a resort village [in Okinawa]
> called Onna, on the island's west coast. The research community in Japan,
> including those who are involved with the project, is largely skeptical, but
> success here could mean a new model for an ailing university system.[1]

Japan is frequently criticized for its rigid and complex bureaucracy, a
culture poorly suited to competing with Western powers, especially in the
arenas of research and technology. The Japanese university system is not
shielded from this bureaucratic and hierarchical complexity. National
ordinances have long established the general framework for establishing
and operating institutions of higher education. Basic elements such as the
university's objectives, organization, terms of study, academic degrees, and
academic staff and their duties, are stipulated in the School Education
Law. On the other hand, academic staff qualifications, frameworks for curricula, graduation requirements, and minimum facility/equipment standards
are prescribed by the University Establishment Standards, an ordinance
from the Ministry of Education, Science, Sports, and Culture.[2]

273

The social culture in the basic sciences in Japan is also mired in outmoded tradition. Bright young Japanese investigators, eager to mount independent and innovative research programs, are often frustrated by a rigid hierarchical system that in many ways rivals the old-fashioned mentality of some European universities. Consequently, many of Japan's brightest and most successful students, who venture to the United States for postdoctoral training and experience the open meritocracy of the American academic system, remain there. But like much of Asia, Japan is awakening to the necessity for administrative reform if it is to remain competitive in an increasingly global society.

In 1992 Tokyo University President Akito Arima, an internationally prominent Japanese physicist and former president of the University of Tokyo, and Science Dean Ikuo Kushiro orchestrated the first outside review of a Japanese institution of higher education, by a committee that included Western scientists—among them Brenner. Not surprisingly, the 65-year old Brenner was invited—implored actually—to spearhead a bold experiment in educational and research reform in Japan, to establish a new and academically open graduate university, the Okinawa Institute of Science and Technology (OIST). Although still in the midst of his adventures in Singapore—and elsewhere—Brenner accepted the invitation.

OIST is the brainchild of Koji Omi, an experienced politician who served in Prime Minister Junichiro Koizumi's cabinet both as Minister of State for Science and Technology Policy and as Minister of State for Okinawa and Northern Territory Affairs. This meld of portfolios was fundamental to cutting through the red tape and political machinations that might otherwise have challenged the development of a modern Western-based graduate university in Japan. A staunch advocate of reform and the author of two books on the imperative of launching and maintaining Western style education and research in Japan, Omi eagerly embraced the Koizumi revolution. Soon after his appointments to the Japanese cabinet, he announced plans to create a new graduate university in Okinawa. In formally endorsing this initiative, the Koizumi cabinet noted: "[E]xisting universities with rigid departmental structures face difficulties in developing new systems to accommodate integrative research and education." The cabinet also stressed its wish to "provide a successful model of a research university for Japanese academia."[3] To that end, the government drew attention to the importance of providing "a fully internationalized environment" and announced that the new Institute "will encourage the use of English as the working language of the graduate university, [in the hope that] the infusion of differing cultural mindsets will develop a creative atmosphere that nurtures an original and questing spirit."[3]

Omi convened an ad hoc advisory committee chaired by Professor Arima and sought out as his special advisor the Japanese physicist G. P. Yeh, a graduate of MIT and an investigator at the Fermi National Accelerator Laboratory (Fermilab) in Batavia, Illinois. The first meeting of Arima's committee was convened in August 2001 and laid the foundations for an International Advisory Committee comprised of prominent scientists in the life sciences and physics, mainly from the United States. The advisory committee held its first meeting in Los Angeles in the spring of 2002 and met again in Okinawa a few months later. The enterprise gained further momentum in January 2003, when Prime Minister Koizumi announced his emphatic support for the Okinawa initiative in his address to the Japanese Diet, Japan's bicameral legislature.

In the same month a third meeting of the International Advisory Committee, now expanded to 20 members (including Brenner), was convened in San Francisco. The committee appointed Jerome Friedman (a Nobel Laureate in physics) as acting-president of the new university and Brenner as vice-president. When Friedman announced his reluctance to serve as president, Minister Omi aggressively lobbied Brenner to accept this responsibility.

Brenner met with Prime Minister Koizumi at his official residence in July 2004. In response to Koizumi's publicly stated hope that Brenner would make the OIST one of world's leading universities, Brenner (modestly) replied: "As a first step I will seek to realize the school the top institute in all of Japan."[4] In March 2005 the Japanese Diet approved the establishment of the Okinawa Institute of Science and Technology Promotion Corporation (OIST PC) and charged this entity with the preparation of the institute for operational use. Brenner was designated the first president of OIST—an unprecedented move in a country long known for shunning outside leadership.

Events progressed rapidly. The town of Onna on Okinawa's beautiful western seaboard was selected for the construction of the institute, with a targeted opening date in the spring of 2009. Meanwhile, temporary facilities at Seaside House (formerly Hakoonso) were erected to accommodate meeting and conference facilities and to house four outstanding research teams led by prominent Japanese scientists, all enthusiastic proponents of change in the Japanese university system. Physicist Arika Tonomura (a foreign member of the U.S. National Academy of Sciences) relocated to Okinawa from Hitachi to work on electron-beam microscopy. Mathematical engineer Kenji Doya of the Advanced Telecommunications Research Institute in Kyoto and neurobiologist Shogo Endo from the Brain Science Institute near Tokyo

established programs in neurobiology, and cellular biologist/geneticist Mitsuhiro Yanagida from Kyoto University established a program on cell cycle control in yeast. In late 2005 the 78-year-old Brenner, with research teams already
thriving in Singapore, Cambridge, and the Salk Institute in La Jolla, California,
established his own program in molecular neuroscience at OIST.

There is no free ride for the Japanese research teams at OIST. Each
team leader was carefully selected in a national competition, and each
research group is subject to periodic peer review by outside reviewers approved by a Board of Governors chaired by Brenner. At the time of this
writing, two of the groups had been reviewed. A committee chaired by the
American neurobiologist and Nobel Laureate Torsten Wiesel reviewed Kenji
Doya's research program. Another chaired by Nobelist Tim Hunt reviewed
Mitsuhiro Yanagida's enterprise. Both programs were approved for funding
for a further five years.

Nor has Brenner himself enjoyed a "free ride." Regardless of the noble
intentions for university reform coming from the highest political office in
Japan, Brenner faced both subtle—and not so subtle—resistance to many
of the directives he issued in his capacity as President of OIST and Chairman
of the Board of Governors of OIST Promotion Corporation (PC). A significant
number of Japanese scientists voiced pessimism and dissent about the
Okinawa experiment and its projected cost (estimated at about $250 million).
The press contributed further fuel to the discord when *Yomiuri Shimbun*, a
prominent Japanese newspaper, accused Brenner of spending too little time
in Okinawa for someone with a full-time salary. Rushing to Brenner's defense,
Kiyoshi Kurokawa, science advisor to Prime Minister Abe, noted in a rebuttal:
"[I]t is simply amazing to consider what he has been doing for us."[5] Kurokawa
also pointed out that Brenner's annual salary was far below that of university
presidents elsewhere in the world. "*Yomiuri* has missed the point and is dwelling on a miniscule technicality," he stated.[5] A seasoned veteran of persuasion
and a skilled negotiator in his own right, Brenner never considered conceding
to the bureaucratic manipulation and intrigue that plagued the emergence of
OIST. He moved pieces around his chessboard, countering one offensive
after another, until his adversaries were politely dismissed, moved to other
positions, or ultimately relented.

The appointment of a science advisor to the prime minister was another
first in Japan, and the choice of Kurokawa for this position reflects the
commitment to following the examples set by western countries. Kurokawa,
who served as professor of medicine at UCLA from 1979 to 1984 before

returning to Japan, is outspoken about the need for educational reform in Japan. Kurokawa told an interviewer from *Science*:

> At the leading [Japanese] universities you have to choose which academic department you are heading toward when taking the entrance exam. Why does it have to be that way? Let high school students study whatever they are interested in and get the universities to allow more flexible choices. We must also internationalize the universities by aiming for 30% of our undergraduates being foreign. And finally, [we] have to reform the Japanese hierarchical academic system in which junior researchers work under department chairs. That destroys the creativity and independence of younger professors. Under [the present] inbred system [we're] just nurturing cloned professors.[6]

In 2005 a Board of Governors chaired by Brenner was established to oversee OIST PC and to advise the university president. Board membership was expanded in 2007 to include foreign notables such as Jerome Friedman and Steven Chu (another Nobel Laureate in Physics and the 12th U.S. Energy Secretary). Jean-Marie Lehn (Nobel Laureate in Chemistry) and Tim Hunt, Susumu Tonegawa, and Torsten Wiesel (Nobel Laureates in Medicine or Physiology) also became members of the Board. The following year OIST delivered on its intention to make the university international. The 2006 Annual Report noted that the number of research units had increased to 13, 6 led by non-Japanese scientists.

By 2007 Brenner was satisfied that OIST had gained sufficient traction to hand the reins of day-to-day management to someone permanently based in Okinawa. He identified another foreign visionary, Robert Baughman. Baughman earned a Ph.D. degree in chemistry at Harvard before moving into neuroscience at Harvard Medical School. A man of considerable administrative talent and energy, Baughman served as a senior research executive in the Institute of Neurology at the NIH, where he honed his administrative skills while introducing new technologies and mastering the ins and outs of technology transfer and the management of intellectual property.[7]

Brenner's choice of Baughman was also motivated by cultural considerations. Baughman was once a visiting lecturer in English at Tokyo Medical and Dental University and at Sapporo Medical School and is thus fluent in Japanese. His Japanese wife Hidemi provides him with valuable insights and guidance and has helped him "understand and appreciate the complex subtleties of Japan's culture."[7] At the time of this writing, Baughman's official titles are vice president/executive director and senior advisor to the president of OIST.

OIST has been appropriately attentive to its relationship to the Prefecture of Okinawa. In a 2007 address to the local community, at an event celebrating the second anniversary of the launching of OIST, Baughman stressed that "OIST must be a productive member of the community in Okinawa or the international effort will not succeed. OIST must build a cooperative relationship with the community in all aspects of the function of the university and I will work to ensure that this is achieved."[8] True to his word, in 2008 mayors from 13 cities and towns in northern and middle Okinawa visited OIST, where plans for the academic and research program, the legal structure, and the surrounding community area were presented.

How the Okinawa experiment, with its intended influence on promoting the integration of foreign scientists, will turn out remains to be seen. However, it is sobering to note that the change in Japanese academia extends well beyond this attempt at academic reform. A 1994 editorial in *Nature* reported,

> Gone are the halcyon days when industrial and university employees accepted their lot in return for benefits such as the security of lifetime employment. ... In the past year, a researcher sued his university when his contract was not renewed [and] a university president was run out of office by faculty members.[9]

Concurrent with the exciting initiative underway in Japan, a parallel transformation was unfolding in the United States at the Howard Hughes Medical Institute. Howard R. Hughes is probably best remembered as an eccentric billionaire who spent the last years of his life in virtual isolation. But Hughes was a man of extraordinary intellect, energy, and diverse talents. He dabbled in movies; designed, constructed, and raced airplanes; built TWA into a premier international airline; and developed the Hughes Aircraft Company into one of the country's largest and most important defense contractors.[10]

Hughes's most enduring accomplishment is probably the creation of the medical institute that now bears his name. His vision of scientific philanthropy was neither modest nor ordinary. He wanted his medical institute to be committed to basic research; to probe "the genesis of life itself."[10] In 1984 a group of trustees assumed responsibility for the institute and reaffirmed its primary mission of basic medical research. The trustees guided the organization through the sale of the Hughes Aircraft Company (which the Institute owned entirely) and the period of rapid expansion that followed. Sustained by the financial resources derived from that sale, the Howard Hughes Medical Institute (HHMI) is now the nation's largest private source of support for biomedical research and science education in the United States, making annual contributions that approach half a billion dollars.[10]

At the end of the 1990s, Nobel Laureate Tom Cech replaced Purnell Chopin as president of HHMI, and another prominent biomedical investigator, Gerald Rubin, was appointed vice-president to succeed Maxwell Cowan. This change in leadership infused an enthusiasm for changing the long-established tradition of funding a large number of HHMI investigators distributed among various academic institutions around the United States. "The HHMI endowment was increasing and we could easily have increased the number of HHMI investigators in the country," Rubin related. "But we didn't think that this was particularly needed, since in 2000 the NIH budget had doubled and good investigators were being funded. So we began to think of more creative ways of spending money."[11] Following extensive and wide-ranging discussions, what ultimately emerged was the decision to establish a new HHMI campus in the countryside of northern Virginia. At the time of this resolution, however, details concerning the nature of the program were far from resolved.

Many ideas were considered for the new campus. The HHMI leadership was not keen to recreate a Salk Institute type of environment, in which investigators are strictly dependent on soft money. This result could readily be achieved by simply enlarging the pool of existing Hughes investigators. Rather, the philosophy leaned toward supporting a program in which scientists are enabled to conduct research with minimum distraction and interference, one unlikely to be found anywhere else in American academia. So a critical question was posed. "What is lacking in the U.S. basic science support structure?"[11] After extensive discussion and consultation, the group agreed to establish a program that would support long-range "blue-sky" research in the mode long practiced at premier research establishments such as the Bell Laboratories, Xerox Park, and IBM. The program was to encourage challenging, high-risk research that carried no guarantee of success; research that is difficult to carry out in American universities for all sorts of well-known historical reasons. The action plan was to build a new facility housing research teams of limited size that would be funded sufficiently to free them from the need for any other form of financial support.[11]

Some argued that the absence of a tenure-accruing system would discourage people from applying for positions. However, the consensus opinion was that the organization was not interested in junior scientists who worried about tenure. Rubin related:

> We wanted bright and innovative young scientists who weren't burdened with the operational aspects of running a lab and who wished to be in the lab working with their own two hands. Basically we wanted to establish a research environment in which one didn't have to worry about funding from the outside, which had outstanding core support, including an outstanding machine shop directed by capable engineers, and in which research teams were of limited size, thus forcing people to collaborate.[11]

Janelia Farm, the site chosen for this grand experiment, was one of Virginia's last examples of the country house ideal that spread from England to America in the late 19th century. Sadly perhaps, the era of genteel country living with live-in servants on large estates could not survive the harsh economic climate of the Great Depression and World War II. Many of the great country estates in America disappeared. Situated within easy distance from the main HHMI campus in Bethesda, Maryland, Janelia Farm occupies extensive acreage in Virginia. Beyond the daunting scientific challenges, this rural location presented a considerable social challenge. Living facilities to accommodate families had to be erected. Young scientists and students eager for occasional distraction from the laboratory have to look to Washington, D.C., Bethesda, and Chevy Chase. However, the place readily lends itself to future expansion, and its tranquil surroundings invite an attractive environment for intellectual endeavors. When the opulent Janelia Farm facility (estimated to have cost about $700 million) opened its doors in October 2006, Rubin assumed its directorship with enthusiasm and resolve.

Rubin had no firsthand experience of the inner workings of facilities such as the Bell Laboratories. But he was intimately familiar with the LMB, having been a graduate student there under Andrew Travers's mentorship. At the LMB he had encountered the formidable Sydney Brenner. Rubin related:

> Sydney sleeps very little so I often had what I would like to say were "conversations" with him at night. But they weren't really conversations; they were more like monologues from him, many delivered between two and three in the morning. Typically I'd go to the coffee room for a drink and would find Sydney there. Since I was the only available audience he would say: "Gerry, sit down and let me tell you about this, that or the other." This really meant that he hadn't had an audience to talk to for the past 5 hours or so and was audience-hungry! So I got to talk with him a lot. He was more of a mentor than a research advisor and in retrospect I got more advice and mentoring from Sydney than the average graduate student in the US did, or does.[11]

Brenner agreed to becoming actively involved in planning and guiding the mission of the new facility. He stressed to Rubin and others the importance of keeping a narrow research focus. "You have to be seen as the place to be for a particular research theme," he advised. He also convinced the leadership that it should, at all costs, avoid the well-worn cliché of a broad mix of cell biologists, biochemists, and molecular biologists, a community of investigators that would inevitably brand the enterprise as nothing more than another private research establishment.

When the plans were generally decided and the endeavor was ready to be launched, Rubin convinced Brenner to serve as one of the first nonresident senior fellows. In this capacity Brenner would visit Janelia Farm several times a year for short periods. At one time Brenner considered launching yet another modest research lab of his own at Janelia Farm. But in a rare practical moment, he realized that in approaching his eighth decade, he had taken on more than enough. He elected instead to become a half-time employee of the HHMI, spending three months a year at Janelia and three months a year in Terry Sejnowski's laboratory at the Salk Institute on the HHMI budget.

In 2003 a series of planning workshops was convened to identify specific areas of research. What emerged were nanotechnology, single cell biochemistry, imaging, membrane proteins, and neurobiology. "I don't claim that these are unique areas of research," Rubin stated. "But the particular mix of scientists we identified is unique. We now have more Ph.D.'s in physics than Ph.D.'s in biology. And in neurobiology we have more people doing computer science than anything else."[11]

Brenner makes concerted efforts to arrange his visits to coincide with one or more of the many conferences now featured on a regular basis at Janelia Farm. When on site Brenner spends much of his time wandering around the hallways and laboratories, talking with anyone and everyone. From the newest graduate students and postdoctoral fellows to the most senior principal investigators, he asks about their ideas and research goals and freely shares his own. "He holds court," Rubin remarked, "which is exactly what I want him to do. He is not at all elitist in this setting. He's just as happy talking to a first year graduate student as he is talking to anyone else. He's a treasure to have around and he provides input into most core decisions."[11]

"Janelia is an interesting experiment," Brenner muses. "It attempts to encourage young scientists to undertake long-term difficult or high-risk projects." Rubin's invitation to serve as a Senior Fellow prompted the aging Brenner to reflect on the uniqueness of the LMB in its hey day. One revelation that came readily to Brenner's mind was the conviction that, regardless of individual areas of scientific expertise, everyone at the LMB was focused on a single biological challenge—DNA and coding. "It didn't matter if you were a crystal-lographer or a geneticist or a chemist. Everyone had the same common goal." Another important realization that surfaced was that "we [members of the LMB community] not only executed the science, we also developed the technologies and equipment that we needed." Thus, Brenner strongly advised Rubin to build connections to the outer constituents of HHMI. Heeding this advice Rubin has initiated a program in which HHMI fellows around the country travel to Janelia to collaborate with in-house scientists. Rubin thoughtfully established a panel of outside HHMI fellows to help select new converts to Janelia.

Rubin understands and appreciates Brenner's inclination to dash around the world, keeping involved with his many efforts.

> At this stage of his career Sydney believes that providing high-level advice is a good way to influence science. It's very satisfying to be able to do this and he truly enjoys it. But he doesn't want to be tied to any one laboratory. He likes to pop in once every so often and examine the data and give fresh directions. I once asked him why he didn't settle down in a single institution that he could call home. His response was: "It's better to have six mistresses than one wife!"[11]

Rubin is excited and optimistic about the future of Janelia Farm. However, he is also acutely aware that it is still an experiment in progress. "We may not get it exactly right at first, but we'll adapt. We'll revise the hypothesis; like any good scientist would do."[11] Brenner shares Rubin's optimism and recalled that

> In the good old days of molecular biology we all worked in small groups. The LMB was nothing more than a large constellation of people operating like that. But I don't think this can happen in the present climate of American academia in which grants are difficult to obtain and having to get tenure—and all that sort of thing, not to mention being projected upward into administration and management. Janelia Farm is the only place in the US that I think has a real chance of bringing about a new revolution in biology at this time. In the UK such notions have simply disappeared.

It is fair to state that all three of Brenner's grand mentoring efforts differ to a degree. The Singapore enterprise represents an excellent example of Brenner's skill in harnessing the energy, commitment, and financial resources of a young and prosperous country eager to join the front rank of international biomedical research. In contrast, OIST provides an educational example of how long-standing cultural influences that are perceived to hinder cutting-edge research can be altered, although the final evaluation is not yet in. Finally, the Janelia Farm experiment is designed to free outstanding young scientists from the onerous burden of writing grants in search of financial resources from funding entities not known for supporting high risk and innovative science. It is also a heroic attempt to bring scientists from multiple scientific disciplines together in the hope of achieving the sort of cross-fertilization that is difficult, if not impossible, to achieve in more structured environments.

27

Enfant Terrible

One should always have a complete sense of how ludicrous one can be

IN 1994 LEWIS WOLPERT CONCLUDED AN INTERVIEW WITH BRENNER, posing a series of questions to provide a sense of how Brenner hoped to be remembered. Asked to comment on his perceptions of his strengths and weaknesses, Brenner denied the arrogance and insensitivity for which others condemn him.

> I don't think I am arrogant or insulting—I think that I'm basically pretty honest. Science is very important and what we're trying to do [in it] is very important. So I sometimes do express impatience with people who lie in the way.... I believe that my good sense of humor, particularly about myself, is one of my great advantages. I think that it's very important to have a developed sense of the ludicrous, especially about oneself; not necessarily how ludicrous one is, but how ludicrous one can be. Pomposity is one of my great fears. I think that pomposity in an old person is bad enough. But pomposity in a young man is absolutely beyond the pale.[1]

Earlier chapters documented Sydney Brenner's quick and often biting wit. This trait—coupled with his irreverence, general disdain for bureaucracy, and any and all imposed rules and regulations—has made him the scourge of university and institutional administrators. But it has also endeared him to generations of friends and colleagues, and to countless audiences of scientists, students, postdoctoral fellows, and others who have crossed his path. The mentality of the enfant terrible (allegedly assigned by François Jacob) has lent Brenner an engaging quality that has softened his formidable scientific genius, a quality that lends him a reassuring sense of being human. As a South African commentator once put it,

> Brenner is the "happy possessor of a 'talent'" [a phrase borrowed from Jean-Paul Sartre] noticeable as these things tend to be from a young age. We might

add, having had the pleasure and first-hand experience of his at times sardonic, at times delightfully dolorous wit, that he is also the "humorous" possessor of talent, cutting always, precisely, to the issue.[2]

Indeed, observing Brenner during the rare moments in conversation when he is listening rather than speaking, one surmises that his silence is merely an interlude to a joke or anecdote appropriate to the conversation at hand. Much of the appeal of Brenner's repertoire of stories stems from the manner in which he uses humor to underscore serious messages. He has been aptly described as "one of biology's mischievous children; the witty trickster who delights in stirring things up. A true scientific original over the past fifty years, Brenner uses his drollness and his facility at solving problems to blaze through one breakthrough after another. . . ."[2]

Brenner's clever wit provides an essential reality check for him, a way of remaining grounded in the midst of the gravity of everyday life. In fact, he professes frank disdain for those who cannot and do not laugh at themselves. His amusing conversational style keeps his audiences engaged—particularly when he is explaining science to lay listeners—which he considers important and at which he is remarkably effective. Consider how he once alerted a lay audience to the folly of expressing rash opinions about the genetics of complex behaviors—such as speech.

> We are about to sequence the chimpanzee genome and some will say that if we subtract the two genomes from each other, we will be left with an extra gene in man, which is the gene for language. And that will be called the Chomsky gene [referring to the celebrated American linguist Noam Chomsky]. But consider an alternative. Maybe what chimpanzees have learnt—learnt in their genomes—is that talking gets you into trouble and they have evolved a language suppressor gene. Of course nobody will allow us, and probably nature doesn't allow us either, to cross chimpanzees and humans to find out which is dominant. But if chimpanzees do indeed have a language suppressor gene, we would have a good name for it: it will be called the Chimpsky gene.[2]

Brenner's fondness for holding forth on anything and everything that captures his diverse interests prompted him to contribute a monthly column to the journal *Current Biology* from 1994 to 2000. These columns, initially titled Loose Ends and later, False Starts, address a wide range of scientifically related topics and activities—some seriously, most frivolously, but all entertainingly written. In some of these, Brenner adopted the pen name Uncle Syd to communicate letters to Willie, an imaginary graduate student/post-doc/faculty member, whom he "mentored" throughout Willie's career.

The notion of a regularly featured column in *Current Biology* was the brainchild of the journal's founder Vitek Tracz. An innovative entrepreneur in the science publishing industry[a] (and a formidable art connoisseur), Tracz sought to publish opinions penned by a scientist of note, someone able to present interesting ideas in an arresting manner. An engaging personality in his own right with a flair for the offbeat, Tracz identified Brenner as his primary choice. To the astonishment of many, Brenner agreed—in exchange for bottles of fine wines. The task of managing the column was the province of senior editor Peter Newmark, formerly with *Nature* magazine. At the outset Newmark anticipated the challenge of hounding Brenner for columns on a regular basis, doubting that someone so busy would submit to the rigor of a monthly deadline. Indeed, when Newark was first informed by Tracz about Brenner's interest, he concluded that this was another bit of wishful thinking squeezed out of an alcoholic lunch, and offered up to him as editor as "a *fait accompli* rather than a *grande illusion*."[3] Colleagues of Sydney's chuckled politely at the notion that he would deliver anything on time and with regularity more than at the prospect of their content.[3]

Brenner proved the skeptics wrong. For the first year or so, Newmark experienced some anxiety as each monthly deadline approached, especially when Brenner's part-time secretary (if she could be reached) had no idea if he intended to deliver a column, or from where.[3] On a few occasions, Newmark resorted to asking May if she had a contact number. But May seldom knew which country he was in! As Sydney did not use e-mail ("I prefer she-mail," he would pun), all Newmark could do was wait by the fax machine as the deadline approached.[3]

Brenner delivered his articles from wherever he happened to be at the time, often faxed from a hotel somewhere, sometimes at the eleventh hour.[3] They were hand-written in his immaculate style, with periodic edits (suggesting there were few, if any, previous drafts), and almost always the precise length specified. The themes Brenner explored for his columns

---

[a] Tracz has spawned and sold (to his considerable financial benefit) various biomedical publishing enterprises including Current Science Ltd., which developed the *Current Opinion* series of journals in areas of clinical medicine and biology, as well as the journals *Current Biology*, *Structure*, and one of the first communities on the web, *BioMedNet*. The sale of *Current Biology* and *BioMedNet* was followed by a number of new publishing ventures, including *Current Science* and *BioMed Central*. See http://www.scienceblog.com/community/older/2001/A/200110977.html.

were entirely of his own choosing, but he and Newmark would confer when he wanted suggestions. Newmark rarely found it necessary to edit Brenner's contributions. "My primary task was," in Newmark's words, "that of removing cumbersome phrases and watching for any words that might invoke suits of libel and defamation."[3] Occasionally, Newmark had to resort to the red pen, but only to add an explanatory phrase, avoid repetition, or re-order the text. Only once did Newmark reject a column outright; Brenner did not complain and immediately wrote a replacement.

Brenner published his first Loose Ends contribution in the fourth volume of *Current Biology* in 1994. The columns were initially featured at the back of each issue. However, when the journal was redesigned in 1998, Brenner retitled the entries False Starts, and they were moved to the front of the journal. In 1997 a collection of the columns were organized in book format. The result, *Loose Ends from Current Biology*, afforded Newmark the opportunity of organizing 38 of the pieces into 6 sections: All the world's a lab; How it was; The seven deadly curs'd sins; Scientifically speaking; Molecular biology by numbers; and Beating the system. Instead of a formal preface, Brenner crafted an introductory piece entitled Loose Beginnings. The compilation, which features most of the Loose Ends pieces from *Current Biology*, is now out of print. However, readers may search for the book from second-hand booksellers on the internet or download the entire collection from the *Current Biology* website.[4]

Pretentious and often confusing descriptors such as *in silico* biology, systems biology, and even bioinformatics, exasperate Brenner. Predictably, more than one of his Loose Ends/False Starts pieces voiced his displeasure with this trend. In a 1998 False Starts column called Net Prophets (a marvelous example of Brenner's penchant for puns) he wrote:

> Statements that "we have come to do biology in a new way" or "there is a new paradigm in biological research" are now commonplace. Nobody seems to be satisfied by a single good experiment that gives a precise answer to a well-formulated question, which was the old way we did biology. On the contrary, there is now a belief that a mass attack on parallel fronts can provide a database of all the information in one concerted effort, and all we need is a computer programme that will give everybody all the knowledge they need. . . . This approach has generated two new areas of activity. One, Bioinformatics, is simply pretentious; the other, Functional Genomics, is ridiculous.

Peer review of manuscripts and grant proposals are other favorite subjects of Brenner's scorn. However, not all his columns are blatantly humorous.

A lengthy follow-up piece on the vexing topic of peer review, pointedly entitled Moron Peer Review, reflected Brenner's own experience when he once (and only once; see Chapter 23) applied for an NIH grant. He is equally disparaging about peer review of manuscripts.

> Some years ago we published a paper in a genetics journal . . . One referee had no complaints, but the other said we should do genetic experiments to prove our point. The Editor's letter urged us to pay attention only to this referee's comments and said the manuscript was seriously defective and could not be published without the genetic experiments. The following telephone conversation then took place:
>
> S.B.: (after introducing the matter): Did you read the paper yourself?
>
> Editor: No, I cannot be expected to read everything that crosses my desk.
>
> S.B.: Are you aware that the referee you selected either can't read English, or more likely, is a total moron? The experiments he asks for were done and published a few years ago. They are clearly referred to in the paper, and the physical evidence supports them.
>
> Editor: (silence)
>
> S.B.: Who is the referee?
>
> Editor: I can't tell you that.
>
> S.B.: You should now accept responsibility for your bad choice and since his comment is both groundless and worthless, I assume you will now accept the paper?
>
> Editor: No, we cannot go back on our original decision; there is no appeal.[5]

Loose Ends is accompanied by a little-known series of delightful cartoons by Andrzej Krauze, orchestrated by *BioMed Central* and featured under the byline: In the words of Sydney Brenner. The caption of my favorite cartoon (see photo section) reads: "The definition of data-mining: what's my data is mine and what's yours is also mine."

Brenner's wit is not always amusing to everyone. As we have seen, he can be cuttingly sarcastic, and, though veiled in humor, critical comments about others sometimes border on the unkind. One of his colleagues stated:

> I have a sense that when Sydney first meets people he either likes or dislikes one. If you're on the "like" list you're a friend of his for life. But if you're on the "dislike" list you have to work really hard to change his mind. I'm not sure what determines his likes and dislikes in relation to

others. Certainly those with pompous or pretentious airs are quickly cut down to size."[6]

Another colleague commented, "Some people take his sarcasm harder than others. But I've never thought of him as being deliberately unkind to anyone."[7] Ultimately, one is left to sort out for oneself the true intention of comments from Brenner such as: "Why sit at home and be miserable, when you can come into the laboratory and work—and be miserable?"[7]

A sensitive soul commenting on Brenner's and Crick's contrasting styles during the question phase of research seminars wrote: "In seminars Crick rarely asked the first question—more often the last. But unlike Brenner's interjections, which to me, often seemed brutal and rude, Crick's questions were always very polite.... "[8] However, there is a more generous, certainly a more thoughtful, side to Brenner. He is frequently consulted by journal editors, publishers, meeting organizers, and others seeking to identify individuals for tasks that carry a measure of distinction besides the promise of hard work. Here he makes considered efforts to select the best and brightest for the task in question, regardless of personal opinions or grudges; and, notably, the recommendation is often made anonymously. Sydney is also known to promote efforts to award well-deserved recognition to colleagues he admires for their contributions as well as their creativity. This, too, is accomplished quietly and without fuss. Enfant terrible is an appropriate descriptor for Sydney Brenner. But sometimes *l'ange de miséricorde* may be more appropriate.

Further in Wolpert's 1994 interview, Brenner addressed his sense of the importance of creativity in science.

> Everybody wants to know how to be creative ... to me the most important thing about creativity is to be unafraid of challenging conventional dogma ... I've been a rebel, or as I was once called, an *enfant terrible*, for a long time. Being a rebel has always appealed to me, largely because I'm convinced that the standard parts of any activity are always petrified at the core. ... Too many people, especially in our Western culture, are brought up to think that daring ideas shouldn't be uttered because they're likely to be wrong. And our culture doesn't promote learning how to turn ideas upside down. It usually imposes the notion of linear thinking. But sometimes it's extremely useful to ask whether in fact an effect is actually a cause—to invert ideas. This is terribly important in science because it's so easy to fall (and stay) in love with one's own theories. Theories should be treated like mistresses. One should never fall in love with them and they should be discarded when the pleasure they provide is over.[1]

"It's very important to daydream about science; to think imaginatively," Brenner continued.

> But to me the essence of creativity in science is to know when and how one can actually prove some of the ideas one daydreams about. So intuitiveness is at the core of a lot of creativity in science; just feeling that one is right about something and going to lengths to actually prove it. This mentality raises a fundamental sort of contradiction. One has to be willing to let go of ideas when they feel wrong, but one also has to be tenaciously obstinate when they feel right. So science asks one to be passionate about invention, but it also asks you to be ruthless and cut off your own hand if it comes to that. There are brilliant people who never accomplish anything. And there are some who have no ideas, but are doers. If one could chimerize them—join them to form a single person—one would have a great scientist. Perhaps that's why science has to operate as a group, as a social unit.[1]

Brenner's influence has extended to all corners of the world. His direct contributions to the growth and development of science in these remote areas, especially Singapore, are also nothing short of spectacular. One is left with little doubt that if post-apartheid South Africa possessed the financial resources to support a serious enterprise in molecular biology, he would willingly lend his services there, too. Brenner's direct contributions to science are matched by his many, but less prominent, indirect contributions. Consider how tirelessly he has worked to influence several generations of scientists, including multiple Nobel Laureates. He has embraced the disciplines of genetics, biochemistry, nucleic acid chemistry, evolutionary biology, developmental biology, and microbiology—and sometimes good old-fashioned common sense. Nor are his scientific contributions limited to intellectual pursuits. He is fearless in developing practical new tools for research in biology, and his influence in biotechnology goes well beyond his collaboration with entrepreneurs and investors. "Sydney loves to invent gadgets and 'chemware' that exploit the logic of life to solve industrial and clinical problems," says Philip Goelet.[9]

Brenner reads widely (one means of coping with the tedium of long flights). With the passing years he has cultivated an increasing interest in biography, history, and in particular, autobiography. "I have come to the conclusion that the past does live on and that it's both interesting and important to read about it." He identified (in 1994) two books of particular interest and relevance to him: the first volume of evolutionary geneticist Richard Goldschmidt's life (*In and Out of the Ivory Tower: The Autobiography of*

Richard Goldschmidt); and an autobiography that Max Born wrote for his family that was eventually published in German (*Mein Leben: die Erinnerungen des Nobelpreisträgers*) and subsequently translated as *My Life: Recollections of a Nobel Laureate.* Both books are about scientists, notably both famous, both Jewish, and both born and raised in a country vilified for violating human rights. Echoes of his own background?

A few reflections on Sydney Brenner the husband, father, stepfather, and grandfather. His devoted wife May suffered ill health much of her life. She passed away in early 2010 at the age of 89, following a protracted illness that kept her bed-ridden for months. During this harrowing period Brenner curtailed his professional responsibilities and obligations as much as possible, tending to May's needs to the best of his ability. While away, he made sure that May was well taken care of by others, notably her eldest son Jonathan, who has a home in Ely. Brenner is also in frequent contact with his biological children and his grandchildren, who conveniently reside in or near London. They view him with obvious fondness, having long since accepted that nothing in the universe can compete with Brenner's passion for his work.

Some who have known Brenner for a long time suggest that at age 83 he is finally mellowing; that he no longer goes out of his way to be confrontational or provocative, at least not in public. Others, perhaps not so close to him, are of the view that he has not changed at all. One sometimes hears Brenner volunteer that he is tired—notwithstanding his capacity to make do with a few hours sleep a night—although such admission hardly translates to slowing down.

This newfound tranquility should not be interpreted as a sign of a diminishing intellectual enthusiasm or interest. Brenner's computer and his many current notebooks contain reams of DNA sequences that he is carefully mining in search of answers to all sorts of burning evolutionary questions. He now sometimes mutters about intentions to write a book—possibly even books. He is especially keen to add further essays to Loose Ends, but he never mentions documenting his memoirs or writing a formal autobiography. While passing years may ultimately relegate Brenner's singular personality to the realm of secondary interest, they will unquestionably grant him a notable place in the history of the natural sciences.

However, Brenner is painfully aware of the debt he owes to chance—the chance of being in the right place at the right time. He comprehends his own considerable scientific successes in the context of having had to learn scientific independence in a remote part of the world with few role models and even fewer mentors. Furthermore, he comprehends the value of arriving in Oxford with very few scientific biases or prejudices, and leaving it with the absolutely certainty that a defining moment in his life occurred one chilly April morning in 1953, viewing the model of the DNA double helix. Asked about his scientific heroes Brenner provided three names: John von Neumann, who he suggested he may envy more than admire; Francis Crick, who according to Brenner always asked the right questions, even though he frequently obtained the wrong answers; and Leo Szilard, "because of his total obliviousness to convention and his complete focus on his own ideas and beliefs."

Reference Sources and Notes*

*Unless otherwise noted all interviews were audio-taped and
are stored, along with most letters, in the Sydney Brenner
Collection at the Cold Spring Harbor Laboratory
Library and Archive*

Prologue

1. Presentation Speech by Professor Urban Lendahl of the Nobel Committee at the Karolinska Institute. See http://nobelprize.org/medicine/laureates/2002/presentation-speech.html.

2. Jacob F. 1988. *The statue within: An autobiography.* Basic Books, New York.

3. Michael Rogers. "The Pandora's box congress." *Rolling Stone*, 19 June 1975, p. 40.

4. Brown A. 2003. *In the beginning was the worm: Finding the secrets of life in a tiny hermaphrodite*, p. 9. Columbia University Press, New York.

5. Hoffenberg R. 2003. Brenner, the worm and the prize. *Clin Med* 3: 285–286.

6. Sydney Brenner's Nobel banquet speech. See http://nobelprize.virtual.museum/nobel_prizes/medicine/laureates/2002/brenner-speech-text.html.

PART 1

Chapter 1

1. Dubb A. 1977. Jewish South Africans: A sociological view of the Johannesburg community. Occasional paper #21, Institute of Social and Economic Research, Rhodes University, South Africa, pp. 19–31.

2. Interview with Phyllis Finn, Johannesburg, South Africa, 17 August 2007.

3. See http://www.gslis.utexas.edu/~landc/fulltext/LandC_34_1_Rochester.pdf.

4. *Die Landstem*, 8 September 1965.

5. Zion M. *Arena*, Wits University, March 2003, p. 40.

---

* Ibid denotes that the reference is the same as the preceding reference; Op cit denotes that the reference is the same as the previously cited reference as indicated in the parentheses.

6. Letter from Sydney Brenner to Lawrence Bragg, 1 February 1966.

7. Letter from Lawrence Bragg to Sydney Brenner, 8 February 1966.

8. Interview with May Brenner, La Jolla, California, 20 January 2007.

9. Wells HG, Huxley J, Wells GP. 1931. *The science of life*. Cassel and Company, Ltd., London.

Chapter 2

1. Murray BK. 1982. *Wits: The early years: A history of the University of the Witwatersrand Johannesburg and its precursors, 1896–1939*, p. 71. University of the Witwatersrand Press, Johannesburg, South Africa.

2. Ibid, p. 93.

3. Dart RA. 1959. *Adventures with the missing link*, pp. 32–33. Harper & Brothers, New York.

4. See http://www.sahistory.org.za/pages/people/bios/roux-e.htm.

5. Phillips H. 2000. What did your university do during apartheid? *J South African Studies* **26**: 173–177.

6. Interview with Phillip V. Tobias, Johannesburg, South Africa, 17 August 2007.

7. Brenner S. "Towards a semantic sociology." *Student Review*, Vol 2, #3, 30 August 1948.

8. Brenner S. "Freedom and the South African universities." *Trek*, October 1949.

9. Tobias PV. 2005. *Into the past: A memoir*, pp. 61–62. Picador Africa and University of the Witwatersrand Press, Johannesburg, South Africa.

10. Letter from Humphrey Raikes to Sydney Brenner, 21 August 1950.

11. Letter from Gerald Blecher to Sydney Brenner, 8 February 1952.

12. Op cit (Note 3), p. 5.

13. Dart R. 1967. *Australopithecus africanus*: The man-ape of South Africa. *Nature* **115**: 195–199.

14. Op cit (Note 1), p. 183.

15. Letter from Sydney Brenner to Raymond A. Dart, 29 November 1982.

16. van Zyl A. 1982. The other side of history: An anecdotal reflection on the political. *South African Medical Journal* **62**: 1008.

Chapter 3

1. Interview with Stanley Glasser, London, 19 July 2008.

2. Interview with Jonathan Balkind, London, 19 July 2008.

3. Interview with May Brenner, La Jolla, California, 20 January 2007.

4. Letter from Harold Daitz to Sydney Brenner, April 1948.

5. Horace Freeland Judson. "Going strong at 75." *The Scientist*, 18 March 2002, p. 6.

6. Letter from Sydney Brenner to Meryl Gray, 4 January 1985.

7. Sydney Brenner, *Theoretical biology*, 7 September 1949, Brenner Collection, Cold Spring Harbor Library and Archive.

Chapter 4

1. Needham J. 1942. *Biochemistry and morphogenesis*. Cambridge University Press, Cambridge, UK.

2. Brachet J. 1944. *Embryologie chimique*. Masson & Cie, Paris.

3. Wilson EB. 1900. *The cell in development and inheritance*. Macmillan, London.

4. Darlington CD. 1937. *Recent advances in cytology*. P. Blakiston's Son & Co., Inc., Philadelphia.

5. Darlington CD, La Cour LF. 1942. *The handling of chromosomes*. Allen and Unwin, London.

6. Brenner S. 1946. The chromosome complement of *Elephantulus*. *S Afr J Med Sci* **11:** Biol Suppl 71.

7. Tobias, PV. *News Bulletin*. Royal Society of South Africa, February 2003.

Chapter 5

1. Hoffenberg R. 2003. Brenner, the worm and the prize. *Clin Med* **3:** 285–286.

2. Letter from Guy Elliot to a colleague, 4 January 1951, Wits University Medical School Library and Archive.

3. Letter from Sydney Brenner to the South African Medical & Dental Council, 20 July 1951.

4. Brachet J. 2005. Reminiscences about nucleic acid cytochemistry and biochemistry. In *The inside story: From DNA to RNA to protein* (ed. J Witkowski et al.). Cold Spring Harbor Laboratory Press, Cold Spring Harbor, NY.

5. Letter from Sydney Brenner to James Norman Davidson, 17 May 1948.

6. Letter from Conrad Waddington to Sydney Brenner, 17 May 1949.

7. Letter from Sydney Brenner to Humphrey Raikes, 20 March 1951.

8. Letter from Humphrey Raikes to Cyril Hinshelwood, 2 October 1951.

9. Letter from Sydney Brenner to Humphrey Raikes, 10 May 1952.

10. Letter from Humphrey Raikes to Sydney Brenner, 12 May 1952.

11. Letter from Cyril Hinshelwood to Sydney Brenner, 29 May 1952.

12. Letter from Sydney Brenner to Cyril Hinshelwood, 10 June 1952.

13. See http://www.absoluteastronomy.com/topics/William_Astbury; see also Bernal, JD. 1963. William Thomas Astbury: 1898–1961. *Biogr Mems Fell R Soc* **9:** 1–35.

14. Caldwell PC, Hinshelwood C. 1950. Some considerations on autosynthesis in bacteria. *J Chem Soc* **4:** 3156–3159; quoted in Crick FHC. 1966. The genetic code: Yesterday, today and tomorrow. *Cold Spring Harbor Symp Quant Biol* **31:** 3–9.

15. Dounce AL. 1952. Duplicating mechanisms for peptide chain and nucleic acid synthesis. *Enzymologia* **15:** 251–258; quoted in Olby R. 2003. Quiet debut for the double helix. *Nature* **421:** 402–405.

16. Judson HF. 1979. *The eighth day of creation: Makers of the revolution in biology.* Simon & Schuster, New York.

17. Interview with May Brenner, La Jolla, California, 20 January 2007.

PART 2

Chapter 6

1. de Chadarevian S. 2002. *Designs for life: Molecular biology after World War II,* p. 41. Cambridge University Press, Cambridge, UK.

2. Ibid, p. 42.

3. Interview with Jonathan Balkind, London, 19 July 2008.

4. Interview with May Brenner, La Jolla, California, February 2007.

5. Telephone interview with Sara Bancroft, January 2009.

6. See http://www.oua.ox.ac.uk/holdings/Halifax%20House%20HH.pdf.

7. Horace Freeland Judson. "Going strong at 75." *The Scientist,* 18 March 2002, p. 16.

8. See http://www.novelguide.com/a/discover/ewb_07/ewb_07_02998.html.

9. See http://www.vanderbilt.edu/exploration/text/index.php?action=view_section&id=1170&story_id=282&images=.

10. See http://profiles.nlm.nih.gov/QL/Views/Exhibit/narrative/genetics.html.

11. Luria SE. 1984. *A slot machine, a broken test tube: An autobiography,* p. 74. Colophon Books/Harper and Row, New York.

12. Brenner S. 1984. Confessions of a biologist. *Nature* **308:** 794.

13. Letter from Humphrey Raikes to Richard Dawkins, undated.

14. Brenner S. 1974. New directions in molecular biology. *Nature* **248:** 785–787.

15. Inglis JR, Sambrook J, Witkowski JA, eds. 2003. *Inspiring science: Jim Watson and the age of DNA,* p. 67. Cold Spring Harbor Laboratory Press, Cold Spring Harbor, NY.

16. Watson JD. 1983. Thirty years of DNA. *Nature* **302:** 651–654.

17. Olby R. 2003. Quiet debut for the double helix. *Nature* **421:** 402–405.

18. Campbell PN, Work TS. 1953. Biosynthesis of proteins. *Nature* **171:** 997–1001.

19. Judson HF. 1979. *The eighth day of creation: Makers of the revolution in biology,* pp. 206–207. Simon & Schuster, New York.

20. Brenner S. 1954. "The physical chemistry of cell processes: A study of bacteriophage resistance in *Escherichia coli,* Strain B12." PhD thesis, Oxford University. Brenner Collection, Cold Spring Harbor Library and Archive.

21. Letter from Sydney Brenner to James D. Watson, 31 August 1953.

22. Brenner S. 2003. The double helix. In *A century of nature: Twenty-one discoveries that changed science and the world* (ed. L Garwin, T Lincoln). University of Chicago Press.

Chapter 7

1. Letter from Humphrey Raikes to Sydney Brenner, 24 November 1953.

2. Letter from Sydney Brenner to Humphrey Raikes, 8 December 1953.

3. Watson JD. 2001. *Genes, girls, and Gamow*, p. xiii–xxiv. Oxford University Press, Oxford.

4. Ibid, p. xxiv.

5. Rhodes R. 1986. *The making of the atomic bomb*, p. 198. Simon & Schuster, New York.

6. Gamow G. 1970. *My world line: An informal autobiography*, p. 122. Viking Adult, New York.

7. Watson JD, Crick FHC. 1953. Genetical implications of the structure of deoxyribonucleic acid. *Nature* **171**: 964–967.

8. Marshall Nirenberg, Lecture at Cold Spring Harbor Laboratory, February 2003.

9. Letter from George Gamow to James D. Watson, 8 July 1953, in Watson JD. 2001. *Genes, girls, and Gamow*, p. 32. Oxford University Press, Oxford.

10. Interview with Paul Berg, Stanford, California, 10 March 2008.

11. Interview with Terry Sjenowski, Salk Institute, California, 13 June 2007.

12. Letter from Sydney Brenner to James D. Watson, 31 July 1954.

13. Judson HF. 1979. *The eighth day of creation: Makers of the revolution in biology*, p. 265. Simon & Schuster, New York.

14. Ibid, p. 332.

15. Brenner S. 1957. On the impossibility of all overlapping triplet codes in information transfer from nucleic acid to proteins. *Proc Natl Acad Sci USA* **43(8)**: 687–694.

16. Heidi Aspaturian interview with Seymour Benzer, 11 September 1990–February 1991, California Institute of Technology Archives, Pasadena, California, p. 16.

17. Ibid, p. 44.

18. Ibid, p. 48.

19. Brenner S, Barnett L. 1959. Genetic and chemical studies of the head protein of bacteriophages T2 and T4. *Brookhaven Symp Biol* **12**: 86–94.

20. Letter from Sydney Brenner to Seymour Benzer, 21 October 1954, quoted in Holmes FL. 2006. *Reconceiving the gene: Seymour Benzer's adventures in phage genetics*, p. 325. Yale University Press, New Haven.

21. Ibid, p. 234.

22. Letter from Sydney Brenner to James D. Watson, 5 July 1954.

23. See http://www.woodshole.com.

24. Letter from James D. Watson to Sydney Brenner, 26 July 1954.

25. Letter from Sydney Brenner to James D. Watson, 31 July 1954.

26. Olby R. 2009. *Francis Crick: Hunter of life's secrets*, p. 264. Cold Spring Harbor Laboratory Press, Cold Spring Harbor, NY.

27. Holmes FL. 2006. *Reconceiving the gene*: *Seymour Benzer's adventures in phage genetics*, p. 279. Yale University Press, New Haven.

28. Letter from Gunther Stent to Sydney Brenner, 6 December 1954.

29. Letter from Sydney Brenner to the Crick family, quoted in Olby R. 2009. *Francis Crick: Hunter of life's secrets*, p. 269. Cold Spring Harbor Laboratory Press, Cold Spring Harbor, NY.

30. Report by Sydney Brenner to the Carnegie Foundation, 12 January 1955.

Chapter 8

1.   Interview with Solly Levin, Johannesburg, South Africa, 17 August 2007.

2.   Letter from Sydney Brenner to Gunther Stent, 27 December 1954.

3.   Letter from Sydney Brenner to James D. Watson, 5 January 1955.

4.   Letter from Sydney Brenner to James D. Watson, 12 April 1955.

5.   Letter from Francis Crick to Sydney Brenner, 12 January 1955.

6.   Letter from Gunther Stent to Sydney Brenner, December 1954.

7.   Letter from Sydney Brenner to Gunther Stent, 27 December 1954.

8.   Letter from Francis Crick to Sydney Brenner, 6 July 1955.

9.   Letter from Francis Crick to Sydney Brenner, 30 August 1955.

10. Letter from Francis Crick to Sydney Brenner, 20 October 1955.

11. Judson HF. 1979. *The eighth day of creation: Makers of the revolution in biology*, p. 312. Simon & Schuster, New York.

12. Letter from Francis Crick to Sydney Brenner, 30 December 1955.

13. Interview with May Brenner, La Jolla, California, 20 January 2007.

14. Ferry G. 2007. *Max Perutz and the secret of life*, p. 174. Chatto and Windus, London.

15. Brenner S. 1957. Genetic control and phenotypic mixing of the absorption cofactor requirement in the bacteriophages T2 and T4. *Virology* 3: 500–574.

16. Report to The National Cancer Association of South Africa, Research Grant #C.S.77/1956.

PART 3

Chapter 9

1.   Interview with May Brenner, La Jolla, California, 20 January 2007.

2.   See http://nobelprize.org/nobel_prizes/medicine/laureates/2002/brenner-autobio.html.

3.   Interview with Jonathan Balkind, London, 18 July 2008.

4. Interview with Carla Brenner, London, 18 July 2008.

5. Judson HF. 1979. *The eighth day of creation: Makers of the revolution in biology*, p. 129. Simon & Schuster, New York.

6. Heidi Aspaturian, interview with Seymour Benzer, 11 September 1990– February 1991, California Institute of Technology Archives, Pasadena, California, p. 51.

7. Holmes FL. 2006. *Reconceiving the gene: Seymour Benzer's adventures in phage genetics*, p. 325. Yale University Press, New Haven.

8. Ibid, pp. 276–277.

9. Letter from George Streisinger to Sydney Brenner, dated both 22 June 1956 and 10 July 1956.

10. Letter from George Streisinger to Sydney Brenner, 25 July 1956.

11. Letter from Francis Crick to Sydney Brenner, 17 July 1956.

12. Brenner S, Horne RW. 1959. A negative staining method for high resolution electron microscopy of viruses. *Biochem Biophys Acta* **34:** 103–109.

13. Op cit (Note 6), p. 63.

14. E-mail from Seymour Benzer to Sydney Brenner, 19 October 2002.

15. Letter from Sydney Brenner to Seymour Benzer, 10 October 1967.

16. Brenner S. 1971. Nonsense mutants and the genetic code: A small piece of molecular genetics. *J Amer Med Assoc* **218:** 1023–1026.

17. Brenner S, Benzer S, Barnett L. 1958. Distribution of proflavin-induced mutations in the genetic fine structure. *Nature* **182:** 983–985.

18. Interview with Jonathan Karn, Cleveland, Ohio, 20 May 2008.

19. Judson HF. 1979. *The eighth day of creation: Makers of the revolution in biology*, p. 347. Simon & Schuster, New York.

20. Lasker Foundation: Albert Lasker Basic Medical Research Award for 1971; Award Description. See http://www.laskerfoundation.org/awards/1971_b_description.htm.

21. Interview with James D. Watson, 13 June 2008.

Chapter 10

1. Pennisi E. 2002. A hothouse of molecular biology. *Science* **300:** 278–282.

2. See http://nobelprize.org/nobel_prizes/chemistry/laureates/1962/perutz-bio.html.

3. Olby R. 2009. *Francis Crick: Hunter of life's secrets*, p. 258. Cold Spring Harbor Laboratory Press, Cold Spring Harbor, NY.

4. Ibid, p. 257.

5. de Chadarevian S. 2002. *Designs for life: Molecular biology after World War II*, p. 285. Cambridge University Press, Cambridge, UK.

6. Max Perutz, *The Medical Research Council Laboratory of Molecular Biology*, see http://nobelprize.org/nobel_prizes/medicine/articles/perutz/index.html.

7. Finch J. 2008. *A Nobel fellow on every floor*, p. 73. The Medical Research Laboratory of Molecular Biology, Cambridge, UK.

8. Ferry G. 2007. *Max Perutz and the secret of life*, p. 174. Chatto and Windus, London.

9. Author e-mail correspondence with Jonathan Karn, 7 January 2009.

10. Op cit (Note 5), p. 272.

11. Op cit (Note 8), p. 177.

12. Tributes from various individuals, Symposium in Honor of Sydney Brenner, 1983, Brenner Collection, Cold Spring Harbor Laboratory Library and Archive.

13. Sulston J, Ferry G. 2002. *The common thread: A story of science, politics, ethics, and the human genome*, p. 36. Joseph Henry Press, Washington, DC.

Chapter 11

1. Hershey A. 1953. Nucleic acid economy in bacteria infected with bacteriophage T2. II. Phage precursor nucleic acid. *J Gen Physiol* **37:** 123.

2. See http://www.alumni.psu.edu/awards/individual/dist_rec2006.volkin.htm.

3. Volkin E. 1995. What was the message? *Trends Biochem Sci* **20:** 206–209.

4. Ibid, pp. 206–209.

5. Judson HF. 1979. *The eighth day of creation: Makers of the revolution in biology*, p. 325. Simon & Schuster, New York.

6. See http://www.ornl.gov/info/ornlreview/v37_3_04/article12.shtml.

7. Jukes TH. 1977. Federation meetings. *Nature* **267:** 8.

8. Astrachan L, Volkin E. 1959. Effects of chloramphenicol on ribonucleic acid metabolism in T2-infected *Escherichia coli*. *Biochem Biophys Acta* **32:** 449–456.

9. Ycas M, Vincent WS. 1960. A ribonucleic acid fraction from yeast related in composition to desoxyribonucleic acid. *Proc Natl Acad Sci (USA)* **46:** 804–811.

10. Op cit (Note 5), p. 348.

11. Op cit (Note 5), p. 402.

12. Jacob F. 1995. *The statue within*, p. 291. Cold Spring Harbor Laboratory Press, Cold Spring Harbor, NY.

13. Ibid, pp. 310–311.

14. Watson JD. 2001. *Genes, girls, and Gamow*, p. 260. Oxford University Press, Oxford.

15. Op cit (Note 12), pp. 311–312.

16. Op cit (Note 5), p. 432.

17. Crick F. 1990. *What mad pursuit: A personal view of scientific discovery*, p. 122. Basic Books, New York.

18. Op cit (Note 12), p. 312.

Chapter 12

1.   Jacob F. 1995. *The statue within*, p. 313. Cold Spring Harbor Laboratory Press, Cold Spring Harbor, NY.

2.   Judson HF. 1979. *The eighth day of creation: Makers of the revolution in biology*, p. 433. Simon & Schuster, New York.

3.   Letter from Sydney Brenner to Matthew Meselson, 7 May 1960.

4.   Op cit (Note 2), p. 437.

5.   Op cit (Note 1), p. 310.

6.   Op cit (Note 2), p. 443.

7.   Op cit (Note 1), p. 317.

8.   Letter from François Jacob to Sydney Brenner, 12 September 1960.

9.   Letter from Sydney Brenner to James D. Watson, 21 September 1960.

10.   Letter from François Jacob to Sydney Brenner, 16 February 1961.

11.   Letter from Sydney Brenner to James D. Watson, 21 February 1961.

12.   James Watson, Nobel Lecture, 11 December 1952. See http://nobelprize.org/ nobel_prizes/medicine/laureates/1962/watson-lecture.html.

13.   Jacob F, Monod J. 1961. Genetic regulatory mechanisms in the synthesis of protein. *J Mol Biol* **3**: 318–356.

14.   Olby R. 2009. *Francis Crick: Hunter of life's secrets*, p. 292. Cold Spring Harbor Laboratory Press, Cold Spring Harbor, NY.

15.   Letter from Max Perutz to Harold Himsworth, 9 February 1961.

Chapter 13

1.   Crick F. 1990. *What mad pursuit: A personal view of scientific discovery*, p. 95. Basic Books, New York.

2.   Crick F. 1963. The recent excitement in the coding problem. *Prog in Nucleic Acids Res* **1**: 213–214.

3.   Wolpert L, Richards A. 1988. *A passion for science*, p. 105. Oxford University Press, Oxford.

4.   Ibid, p. 106.

5.   Horace Freeland Judson. "Going strong at 75." *The Scientist*, 18 March 2002, p. 16.

6.   Ridley M. 2006. *Francis Crick: Discoverer of the genetic code*, p. 103. Eminent Lives Series, London.

7.   Olby R. 2009. *Francis Crick: Hunter of life's secrets*, p. 270. Cold Spring Harbor Laboratory Press, Cold Spring Harbor, NY.

8.   Ibid, p. 227.

9.   Holliday R. 2003. The early years of molecular biology: Personal reflections. *Notes Rec R Soc Lond* **57**: 195–208.

10.   Interview with Muriel Wigby, Stapelford, Cambridge, UK, 11 October 2007.

11. Crick FHC, Barnett L, Brenner S, Watts-Tobin RJ. 1961. General nature of the genetic code for proteins. *Nature* **192:** 1227–1232.

12. Barnett L, Brenner S, Crick FHC, Shulman RG, Watts-Tobin RJ. 1967. Phase shift and other mutants in the first part of the *rIIB* cistron of bacteriophage T4. *Phil Trans R Soc Lond B* **252:** 487–560.

13. Op cit (Note 1), p. 135.

Chapter 14

1. Sarabhai AS, Stretton AOW, Brenner S. 1964. Co-linearity of the gene with the polypeptide chain. *Nature* **201:** 13–17.

2. Yanofsky C. 1963. Amino acid replacements associated with mutation and recombination in the A gene and their relationship to in vitro coding data. In *Synthesis and Structure of Macromolecules, Cold Spring Harbor Symp Quant Biol* **28:** 581–588.

3. Yanofsky C, Carlton BC, Guest JR, Helinski DR, Henning U. 1964. On the colinearity of gene structure and protein structure. *Proc Natl Acad Sci (USA)* **51(2):** 266–272.

4. Op cit (Note 1), p. 13.

5. Interview with Jonathan Karn, Cleveland, Ohio, 20 May 2008.

6. Letter from Sydney Brenner to James D. Watson, 26 August 1962.

7. Neuberger MS, Di Noia JM, Beale RCL, Williams GT, Yang Z, Rada C. 2005. Somatic hypermutation at A:T pairs: Polymerase error versus dUTP incorporation. *Nat Rev Immunol* **5:** 171–178.

8. Letter from Francis Crick to Sydney Brenner, 24 April 1963.

Chapter 15

1. Sydney Brenner Collection, Cold Spring Harbor Laboratory Library and Archive.

2. Brown A. 2003. *In the beginning was the worm*: *Finding the secrets of life in a tiny hermaphrodite*, p. 18. Columbia University Press, New York.

3. Ibid, p. 19.

4. Letter from Sydney Brenner to Max Perutz, 26 September 1963.

5. See http://elegans.swmed.edu/Sydney.html.

6. Smith CUM. 2005. Origins of molecular neurobiology: The role of the physicists. *J Hist Neurosci* **14:** 214–229.

7. Op cit (Note 2), p. 28.

8. Letter from Sydney Brenner to Ellsworth Dougherty, 11 October 1963.

9. Op cit (Note 2), Chapter 2.

10. Op cit (Note 2), p. 25.

Chapter 16

1. Brown A. 2003. *In the beginning was the worm: Finding the secrets of life in a tiny hermaphrodite*, p. 30. Columbia University Press, New York.

2. Ibid, p. 30.

3. Ibid, pp. 27–28.

4. Brenner S. 1974. The genetics of *Caenorhabditis elegans. Genetics* **77:** 71–94.

5. White J. 2000. Worm tales. *Int J Dev Biol* **44:** 39–42.

6. Op cit (Note 1), p. 81.

7. Op cit (Note 1), p. 64.

8. Op cit (Note 1), p. 55.

Chapter 17

1. Letter from Sydney Brenner to John W. Cornforth, 21 November 1968.

2. Brenner S. 1974. The genetics of *Caenorhabditis elegans. Genetics* **77:** 71–94.

3. White JG, Southgate E, Thomson JN, Brenner S. 1986. The structure of the nervous system of the nematode *Caenorhabditis elegans*: The mind of a worm. *Phil Trans R Soc Lond* **314:** 1–340.

4. "The wise man of Janelia." *Howard Hughes Medical Institute Bulletin*, Vol. 20, No. 3, August 2007. See http://www.hhmi.org/bulletin/aug2007/features/wiseman3.html.

5. Brown A. 2003. *In the beginning was the worm: Finding the secrets of life in a tiny hermaphrodite*, p. 95. Columbia University Press, New York.

6. Interview with John Sulston, Cambridge, UK, 10 October 2007.

7. Sulston J, Ferry G. 2002. *The common thread: A story of science, politics, ethics, and the human genome*, pp. 29–30. Joseph Henry Press, Washington, DC.

8. Interview with Robert Horvitz, Massachusetts Institute of Technology, 24 June 2008.

9. Sulston JE, Horvitz HR. 1977. Post-embryonic cell lineages of the nematode, *Caenorhabditis elegans. Dev Biol* **56:** 110–156.

10. Op cit (Note 7), p. 36.

Chapter 18

1. Interview with Paul Berg, Stanford University, 27 April 2007.

2. Editorial. 1974. *Nature (London)* **250:** 175.

3. Evidence for the Ashby Working Party, 26 September 1974, Brenner Collection, Cold Spring Harbor Laboratory Library and Archive.

4. Rogers M. 1977. *Biohazard: The struggle to control recombinant DNA experiments, the most promising (and most threatening) scientific research ever undertaken*, pp. 65–66. AA. Knopf, New York.

5. Horace Freeland Judson. "Going strong at 75." *The Scientist*, 18 March 2002, p. 16.

6. Interview with David Baltimore, California Institute of Technology, 23 July 2007.

7. Letter from Sydney Brenner to Max Perutz, undated.

8. News. 1978. Genetic manipulation: new guidelines for UK. *Nature* **276:** 104–108.

Chapter 19

1. MacLeod AR, Waterson RH, Fishpool RM, Brenner S. 1977. Identification of the structural gene for a myosin heavy-chain in *Caenorhabditis elegans*. *J Mol Biol* **114:** 133–140.

2. Karn J, Brenner S, Barnett L, Cesareni G. 1980. Novel bacteriophage lambda cloning vector. *Proc Natl Acad Sci (USA)* **77:** 5172–5176.

3. Voordouw G, Walker JE, Brenner S. 1985. Cloning of the gene encoding the hydrogenase from *Desulfovibrio vulgaris* and determination of the NH_2-terminal sequence. *Eur J Biochem* **148:** 509–514.

4. Voordouw G, Brenner S. 1985. Nucleotide sequence of the gene encoding the hydrogenase from *Desulfovibrio vulgaris*. *Eur J Biochem* **148:** 515–520.

5. Roberts AN, Hudson GS, Brenner S. 1985. An erythromycin-resistance gene from an erythromycin-producing strain of *Arthrobacter* sp. *Gene* **35:** 259–270.

6. Sulston J, Ferry G. 2002. *The common thread: A story of science, politics, ethics, and the human genome*, p. 24. Joseph Henry Press, Washington, DC.

7. Coulson A, Sulston J, Brenner S, Karn J. 1986. Toward a physical map of the genome of the nematode *Caenorhabditis elegans*. *Proc Natl Acad Sci* **83:** 7281–7285.

8. Interview with Robert Horvitz, Massachusetts Institute of Technology, 24 June 2008.

9. Interview with John Sulston, Cambridge, UK, 10 October 2007.

10. Vines G. "Mapping genes: The bottom-up approach." *New Scientist*, 5 March 1987, p. 10.

11. John Sulston, Alan Coulson, and Sydney Brenner, letter to the editor entitled Genomes. *New Scientist*, 19 March 1987, p. 64.

12. Letter from John Sulston to Sydney Brenner, 10 March 1987.

13. Letter from John Sulston and Alan Coulson to Sydney Brenner, 10 March 1987.

14. Letter from Sydney Brenner to John Sulston and Alan Coulson, 23 March 1987.

15. See http://www.sanger.ac.uk/Projects/C_elegans.

16. Chalfie M. 1998. Genome sequencing: The worm revealed. *Nature* **396:** 620–621.

17. Brenner S. 1973. The genetics of behaviour. *Brit Med Bull* **29:** 269–271.

18. Lewin R. 1984. Why is development so illogical? *Science* **224:** 1327–1329.

19. Letter from James Crow to Sydney Brenner, 21 June 1984.

20. Letter from Sydney Brenner to James Crow, 5 July 1984.

21. Brown A. 2003. *In the beginning was the worm: Finding the secrets of life in a tiny hermaphrodite*, p. 27. Columbia University Press, New York.

22. E-mail from Sydney Brenner to Seymour Benzer, 22 October 2002.

Chapter 20

1. Hargittai I, Hargittai M. 2006. *Candid science 6: More conversations with famous scientists (Candid Science) (Pt. 6)*, p. 34. Imperial College Press, London.

2. de Chadarevian S. 2002. *Designs for life: Molecular biology after World War II*, p. 336. Cambridge University Press, Cambridge, UK.

3. Ibid, p. 342.

4. Ferry G. 2007. *Max Perutz and the secret of life*, p. 209. Chatto and Windus, London.

5. Interview with Walter Bodmer, Stanford, California, 18 May 2007.

6. Announcement. 1977. Brenner's appointment. *Nature* **266:** 768.

7. Interview with Richard Henderson, Laboratory of Molecular Biology, Cambridge, UK, 10 October 2007.

8. Letter from Francis Crick to Sydney Brenner, 6 November 1979.

9. Op cit (Note 7).

10. Interview with Bronwen Loder, Cambridge, UK, 18 July 2008.

11. Letter from Sydney Brenner to David James, Medical Research Council, 5 October 1983.

12. Letter from Sydney Brenner to James Gowans, 24 September 1984, marked PERSONAL. Laboratory of Molecular Biology Archives, Cambridge, UK.

13. Letter from Sydney Brenner to Denis Noble, Medical Research Council, 31 July 1984, Laboratory of Molecular Biology Archives, Cambridge, UK.

Chapter 21

1. Interview with Bronwen Loder, Cambridge, UK, 18 July 2008.

2. Letter from James Gowans to Bronwen Loder, 22 September 1985.

3. Privatization in the United Kingdom Under the Thatcher Government. See http://www.sjsu.edu/faulty/watkins/privUK.htm.

4. Letter from James Gowans to Medical Research Council members, 29 October 1985.

5. Editorial announcement. 1985. Brenner to quit MRC post. *Nature* **318:** 98.

6. Letter from Sydney Brenner to Lewis Wolpert, 19 June 1986.

7. Letter from Sydney Brenner to James Gowans, 19 June 1986.

8. Interview with Keith Peters, London, 18 July 2008.

9. Letter from Sydney Brenner to Benno C. Schmidt, 9 February 1977.

10. Interview with Sydney Brenner. 2009. *Nature Rev Mol Cell Biol* **9:** 8–9.

11. Announcement. 1986. *Nature* **322:** 397.

12. Letter from Sydney Brenner to Aaron Klug, 18 July 1986.

13. Letter from James Alwen to Sydney Brenner, 21 August 1986.

14. Letter from Sydney Brenner to Medical Research Council, 29 January 1990.

15. Medical Research Council document 90/B033; Request for extensions of appointment for two years beyond normal retirement, dated March 1990.

16. Unconfirmed extract from the minutes of the Cell Board held on 15 January 1991; accompanying letter from Michael Kemp to Sydney Brenner, 4 February 1991.

17. Letter from Dai Rees to Sydney Brenner, 24 June 1991.

18. Letter from Sydney Brenner to Dr. Rosie Ward, 1 June 1992.

19. Interview with Jonathan Karn, Cleveland, Ohio, 30 May 2008.

PART 5

Chapter 22

1. Frederickson EH. 2007. *A century of science publishing: A collection of essays*, p. 141. Learned Publishing, Association of Learned and Professional Society Publishers, West Sussex, UK.

2. Interview with Jonathan Karn, 30 May 2008.

3. Brenner S (ed.). 1989. *Molecular biology: A selection of papers*. Academic Press, London.

4. Interview with Keith Peters, London, 18 July 2008.

5. Interview with Philip Goelet, Baltimore, Maryland, 20 May 2008.

6. Rose K. 2003. *Elusive Rothschild: The life of Victor, Third Baron, Rothschild*, pp. ix–x. Wiedenfeld & Nicholson, London.

7. Letter from Lord Rothschild to Sydney Brenner, January 1981.

8. Letter from Sydney Brenner to David Leathers, 9 February 1984.

Chapter 23

1. Letter from Sydney Brenner to Mr. Turner, 25 June 1970.

2. Brenner S, Roberts RJ. 2007. Save your notes, drafts and printouts: Today's work is tomorrow's history. *Nature* **446:** 275.

3. Interview with Phillip V. Tobias, Johannesburg, South Africa, 17 August 2007.

4. Letter from Sydney Brenner to Maurice Wilkins, 6 April 1970.

5. Letter from Sydney Brenner to Princeton University Press, 19 January 1979.

6. Letter from Sydney Brenner to Morris Brown, 6 March 1995.

7. Pennisi E. 2001. The human genome. *Science* **291:** 1182–1188.

8. Interview with Keith Peters, London, 18 July 2008.

9.   Brenner S, Elgar G, Sanford R, Macrae A, Venkatesh B, Aparicio S. 1993. Characterization of the pufferfish (Fugu) genome as a compact model vertebrate genome. *Nature* **366:** 265–268.

10. Interview with Sam Aparicio, Cambridge, UK, 27 September 2008.

11. See http://www.admin.cam.ac.uk/news/press/dpp/2001102601.

12. Aparicio S, Chapman J, Stupka E, Putnam N, Chia JM, Dehal P, Christoffels A, Rash S, Hoon S, Smit A, et al. 2002. Whole-genome shotgun assembly and analysis of the genome of *Fugu rubripes. Science* **297:** 1301–1310.

Chapter 24

1.   Steve Chapple. "Death and how to avoid it." *Los Angeles Times*, 23 January 2006.

2.   See http://www.salk.edu/about/history/html.

3.   Letter from Francis Crick to Sydney Brenner, 22 January 1981.

4.   Letter from Roger Guillemin to Sydney Brenner, 5 September 1985.

5.   Letter from Sydney Brenner to Roger Guillemin, 16 September 1985.

6.   Interview with Richard Lerner, Scripps Research Institute, La Jolla, California, 1 May 2007.

7.   Letter from Michael Kemp to Sydney Brenner, 12 February 1991.

8.   Letter from Sydney Brenner to Michael Kemp, 18 February 1991.

9.   Nicholas Wade. "Scientist at work: Sydney Brenner; A founder of modern biology shapes the genome era, too." *New York Times*, 7 March 2000.

10. Interview with Kim Janda, Scripps Research Institute, La Jolla, California, 1 May 2007.

11. See http://roswell.tobaccodocuments.org/about_TI.htm.

12. See http://www.sourcewatch.org/index.php?title=Tobacco_Institute.

13. Letter from Richard Lerner to Charles Wall, 30 June 1994.

14. Alison Motluck. "Warning shots over 'health hazard' labs." *New Scientist*, 31 August 1996.

15. Document: Proposal for the Creation of the Philip Morris Institute for Molecular Sciences, January 1995, Brenner Collection, Cold Spring Harbor Laboratory Archive.

16. Cohen J. 1996. Philip Morris gives institute a head start. *Science* **272:** 489.

17. Interview with Roger Brent, Molecular Sciences Institute, Berkeley, California, 24 July 2007.

18. Bradford RA. Winter 2003. A man, a worm, and a Nobel. *Salk Signals, A Publication of the Salk Institute* **5(2):** 16.

19. Interview with Philip Goelet, Baltimore, Maryland, 20 May 2008.

20. Phone interview with Sam Eletr, 10 October 2008.

21. 1996. Interview with Sydney Brenner: The world of genome projects. *BioEssays* **18:** 1039–1042.

22. Brenner S. 2002. Life sentences: Ontology recapitulates philology. *Genome Biol* **3**: COMMENT1006.

23. See http://c250.columbia.edu/genomes.

24. Brenner S. 2010. Sequences and consequences. *Phil Trans R Soc Lond B Biol Sci* **365**: 207–212.

25. See http://en.wikipedia.org/wiki/Duhem-Quine_thesis.

26. Hoffenberg R. 2003. Brenner, the worm and the prize. *Clin Med* **3**: 285–286.

Chapter 25

1. Yew LK. 2000. *From third world to first: The singapore story 1965-2000*, p. xiv. Harper Collins, New York.

2. Letter from Victor Rothschild to Sydney Brenner, 4 September 1981.

3. Sydney Brenner notes, "Overview of Biotechnology in Industry," 14–18 November 1983, Singapore, Brenner Collection, Cold Spring Harbor Laboratory Library and Archive.

4. Letter from Sydney Brenner to Dr. Tan Kok Kheng, 12 February 1985.

5. Telephone interview with Chris Tan, 2 March 2007.

6. See http://www.alumni.hbs.edu/bulletin/2001/October/yeo.html.

7. Interview with David Lane, Singapore, 26 April 2008.

8. Interview with Philip Yeo, Singapore, 26 April 2008.

9. Frith M. 2003. Sprawling Biopolis jazzes up Singapore's science scene. *Nature Med* **9**: 1440.

10. Cyranoski D. 2005. Singapore: An irresistible force. *Nature* **436**: 767–768.

11. Interview with Nancy Jenkins and Neal Copeland, 26 April 2008.

12. Interview with Byrappa Venkatesh, Singapore, 26 April 2008.

13. See http://tkmaia.multiply.com/journal/item/253/Sydney_Brenner_Science_Catalyst_of_the_New_World.

Chapter 26

1. Cyranoski D. 2004. Will creativity thrive in an island paradise? *Nature* **429**: 220–221.

2. Osaki H. 1997. The structure of university administration in Japan. *Higher Education* **34**: 151–163.

3. See http://www.youtube.com/watch?v=h4yP_Ltc97E.

4. See http://helpplan.tripod.com/20010622OkinawaNewTechUniv.htm.

5. Newsmakers. 2006. *Science* **314**: 1665.

6. Normile D. 2007. Japan picks up the 'Innovation' mantra. *Science* **316**: 186.

7. Phone interview with Robert Baughman, 16 October 2008.

8. See http://www.oist.jp/en/newsevent/eventcalendar/event-report.html?start=5.

9. Cyranoski D, Chou I-H. 2004. Winds of change blow away the cobwebs on campus: Japan's hidebound university system is being reformed. *Nature* **429:** 210.

10. See http://www.hhmi.org/about/index.html.

11. Interview with Gerald Rubin, Janelia Farm, 20 May 2008.

Chapter 27

1. Brenner S. 2001. *Sydney Brenner: A life in science,* as told to Lewis Wolpert; edited interview with additional material by Errol C. Friedberg and Eleanor Lawrence. BioMed Central, London.

2. James W. 2004. Introduction. In *South Africa's Nobel laureates* (K Asmal, D Chidester, W James, eds.), p. 252. Jonathan Ball Publishers, Johannesburg, South Africa.

3. Phone interview with Peter Newmark, 5 November 2007.

4. Sydney Brenner's Loose Ends and False Starts. See http://www.cell.com/ current-biology/Brenner.

5. Brenner, S. 1996. Loose ends: Out of print. *Curr Biol* **6:** 622.

6. Interview with Paul Wassarman, New York, 14 June 2008.

7. E-mail from Jonathan Karn to author, 9 January 2009.

8. Hogan B. 2005. Rubbing shoulders with giants, in *Genome Life #14,* p. 5. Duke Institute for Genome Sciences and Policy.

9. Interview with Philip Goelet, Baltimore, Maryland, 20 May 2008.

Index

Smith Kline Beckman (pharmaceutical
    company), 220
Smuts, Field Marshall, 35
Smuts, Jan Christian (Prime Minister of
    South Africa), 16
Snow, C. P., 59
"So Much 'Junk' DNA in Our Genome"
    (Ohno), 240
Social responsibility, Brenner's view on, 189
Society for General Microbiology (1960
    symposium), 123
Solexa, Incorporated, 257
Söll, Dieter, 183
Somatic hypermutation, 154
South Africa
    apartheid, 16, 17, 18
    Brenner's 1952 departure from, 54–55
    Brenner's 1956 departure from, 95–97
    Brenner's 1954 return to, 89–90, 91
    Brenner's childhood in, 6–13
    Brenner's Lecturer in Physiology
        position, 91–97
    civil unrest, 15–16
    educational system of, 6, 8, 15
    honors and awards to Brenner, 97
    Jewish immigration, 3–5
South African Communist Party, 17, 18
South African Institute for Medical Research,
    17, 91–92
South African Journal of Medical Science, 40,
    48
South African Medical and Dental Council, 47
South African Nationalist Party, 21
South African Party, 16
South African School of Mines and
    Technology, 16
South African Society for the Advancement
    of Science, 39
South African Students Organization
    (SASO), 19
Southgate, Eileen, "The Structure of the
    Nervous System of the Nematode
    Caenorhabditis elegans: The Mind
    of the Worm," 174
Speaking ability, of Brenner, 235
Speaking style, of Sydney Brenner, 286
Special Branch, 18
Spontaneous mutations, non-revertibility of,
    109
SRI (Scripps Research Institute), 250, 251
Stahl, Frank, 130
Starling, Ernest, 23
Statistical genomics, 241
The Statue Within (Jacob), 123–126
Steinberg, Charley, 150

Stent, Gunther
    bacteriophage work, 64
    Brenner's 1954 communications with, 92, 93
    Brenner's 1954 visit with, 88–89
    offer of position to Brenner at Berkeley
        to, 94
Stoker, Michael, 94
Stop codons, 150–152
Strangeways Laboratory, 161
Streisinger, George
    at Cavendish Laboratory, 105
    colinearity problem and, 104
    communication with Brenner (1956), 104
    food and, 106–107
    host range mutants, 103
Structural Studies, LMB division, 114, 210
"The Structure of the Nervous System of the
    Nematode Caenorhabditis elegans:
    The Mind of the Worm" (Brenner,
    Thomson, Southgate, and White),
    174
Student protest, in South Africa, 18–22
Student Representative Council (SRC),
    Brenner's service on, 19, 20, 22
Student Review - A Journal for Liberals
    (student publication), 19
Students
    Brenner's advice to 17-year-old (1970),
        233–234
    Brenner's interaction with, 179
    lectures by Brenner at Cambridge
        University, 230
    in Singapore, 270–271
Sturtevant, Alfred, 70
Sub-Committee on the Validation of Safe
    Vectors, GMAC, 190
Sulston, John
    on Brenner's supervision style, 179
    C. elegans genome research, 194–199
    cell lineage studies in C. elegans, 175–176
    The Common Thread, 198
    Director of the Wellcome Trust Sanger
        Institute, 199
    election as Fellow of the Royal Society,
        195
    fallout with Brenner, 194–198
    human genome sequencing and, 198, 239
    joining Brenner's C. elegans project, 172,
        174, 175
    knighthood, 199
    letter of intent to Brenner, 197
    Nobel Prize (2002), 116, 178, 201
    "Post-embryonic Cell Lineages of the
        Nematode Caenorhabditis
        elegans" (1977), 177